A Framework for Critical Transnational Research

By foregrounding successful transnational research projects conducted across Latin America and Europe, this edited collection contests epistemological hegemony and heterogeneity in the academy and highlights feasible models for research cooperation across diverse languages, cultures, and epistemologies.

Chapters focus on the practical and theoretical tenets of responsible intra-national research and propose the "Glocacademia" framework as a means of enhancing critical reflection on issues that can inhibit plurilingual, intercultural, and inter-epistemic research. The text offers key recommendations to support institutions and researchers to develop intercultural awareness, multi-level citizenship, and a readiness to embrace diverse knowledge ecologies. The book builds on existing discussions on multiculturalism, interculturality, and transculturality to offer high academic value to the discussion of higher education and research.

Offering important contributions to the study of global academic research, this volume will be of interest to scholars and researchers with an interest in international and comparative education, as well as multicultural studies in education research.

Manuela Guilherme is a Senior Researcher at the Center for Social Studies, University of Coimbra, Portugal.

Routledge Research in International and Comparative Education

This is a series that offers a global platform to engage scholars in continuous academic debate on key challenges and the latest thinking on issues in the fast-growing field of International and Comparative Education.

Titles in the series include:

Primary Mathematics Pedagogy at the Intersection of Education Reform, Policy, and Culture
Comparative Insights from Ghana, Singapore, and the US
Sarah Murray and Princess Allotey

Interpreting International Education
In Honor of Professor Jeff Thompson
Edited by Mary Hayden

The Evolution of Transnational Education
Pathways, Globalization and Emerging Trends
Edited by Christopher Hill, Judith Lamie, and Tim Gore

Belonging in Changing Educational Spaces
Negotiating Global, Transnational, and Neoliberal Dynamics
Edited by Karen Monkman and Ann Frkovich

A Framework for Critical Transnational Research
Advancing Plurilingual, Intercultural, and Inter-epistemic Collaboration in the Academy
Edited by Manuela Guilherme

For more information about this series, please visit: www.routledge.com/Routledge-Research-in-International-and-Comparative-Education/book-series/RRICE

A Framework for Critical Transnational Research

Advancing Plurilingual, Intercultural, and Inter-epistemic Collaboration in the Academy

Edited by Manuela Guilherme

Routledge
Taylor & Francis Group
NEW YORK AND LONDON

First published 2022
by Routledge
605 Third Avenue, New York, NY 10158

and by Routledge
4 Park Square, Milton Park, Abingdon, Oxon, OX14 4RN

Routledge is an imprint of the Taylor & Francis Group, an informa business

Library of Congress Cataloging-in-Publication Data
A catalog record for this title has been requested

ISBN: 9781032127026 (hbk)
ISBN: 9781032127064 (pbk)
ISBN: 9781003225812 (ebk)

DOI: 10.4324/9781003225812

Typeset in Sabon
by KnowledgeWorks Global Ltd.

Também devo a perspectiva do Sul presente neste livro à minha infância, cheia de desafios, nas margens do rio Zambeze, Moçambique, que foi o meu primeiro projecto de investigação sobre plurilinguismo e epistemologias interculturais.

I also owe the perspective of the South present in this book to my challenging childhood on the shores of the Zambeze river, Mozambique, which was my first research project on plurilingualism and intercultural epistemologies.

A minha aprendizagem do mundo, eu dedico aos meus pais, Virgilio e Maria Fernanda, de quem eu aprendi, e aos meus queridos netos, Diego, Rafael e Max, a quem eu gostaria de um dia inspirar.

I dedicate my world learning to my parents, Virgilio and Maria Fernanda, from whom I learned, and to my beloved grandchildren, Diego, Rafael and Max, whom I would like some day to inspire.

Contents

List of Contributors

António Magalhães is Professor at the Faculty of Psychology and Education Sciences of the University of Porto, Portugal. He is also director of research at the Centre for Research in Higher Education Policies. His academic interests focus on education policy analysis, namely on European higher education governance. He has been developing national and international projects in the field of research on higher education. He has written and coedited numerous books and published articles in major journals in the field, including *Higher Education Policy*, *Higher Education* and *Studies in Higher Education*.

Amélia Veiga is Assistant Professor at the Faculty of Psychology and Education Sciences at the University of Porto, Portugal, and researcher at the Centre for Research and Intervention in Education and at the Centre for Research in Higher Education Policies. Her academic interests focus on education policy analysis, namely on the Bologna process and higher education. She has been involved in national and international projects. She has written book chapters and coedited books and published articles on European integration and higher education governance in key journals such as *Higher Education*, *Studies in Higher Education* and *Higher Education Quarterly*.

Manuela Guilherme is a senior researcher at the Centro de Estudos Sociais, Universidade de Coimbra. She was a Marie-Sklodowska Curie grantee (2014–2017). She was awarded a PhD degree in Social Sciences/Education by the University of Durham, UK in 2000 and, in 2001, she was awarded, in Washington D.C., the Birkmaier Award for doctoral research by the ACTFL and The Modern Language Journal. Her academic merit was also recognized in a biographical chapter by Professor Alison Phipps (2012) in The Encyclopaedia of Applied Linguistics, Blackwell Publ. She has coordinated and co-coordinated several international projects funded by the European Commission and has published internationally.

António Teodoro is professor of Sociology of Education and Comparative Education in Lusofona University, at Lisbon, and director of the Interdisciplinary Research Centre for Education and Development (CeiED). Founder and chair of the Portuguese Society of Comparative Education (SPCE-SEC), and member of the Executive Committee and chair of the Constitutional Standing Committee of World Council of Comparative Education Societies (WCCES). Editor of Revista Lusófona de Educação (Lusofona Journal of Education), and member of the editorial board of dozens of journals. Recent book in Routledge: *Contesting the Global Development of Sustainable and Inclusive Education. Education Reform and the Challenges of Neoliberal Globalization* (2020).

Marta Maria Assumpção-Rodrigues is a Psychoanalist, Translator and Professor of Public Policies, University of Sao Paulo; Ph.D. Department of Government and International Studies, University of Notre Dame; M.A. Kroc Institute for International Peace Studies, University of Notre Dame; B.A. Philosophy, Faculty of Philosophy, Languages and Literature, and Human Sciences (FFLCH) of the University of Sao Paulo (USP); B.A. Social and Political Sciences, School of Sociology and Political Science of Sao Paulo Foundation.

José Veríssimo Romão Netto is a Political Scientist, PhD, with experience in structures of government, public policies, and citizen deliberation tools. Holds a postdoc at the University of São Paulo with a postdoctoral internship at the Institute for Local Government Studies at the University of Birmingham - UK. Researcher and consultant for public and private, national, and international organizations. Consultant at Fundação Instituto de Administração (FIA) working on major projects for the Executive and Legislative branches. Advisor in the Graduate Program in Public Policy Management at the School of Arts, Sciences and Humanities of the University of São Paulo.

Suzana Alice Cardoso

Professor Emerita at the Universidade Federal da Bahia, Postgraduate Programme of Language and Culture. PhD in Letters (Letras Vernáculas) from the Universidade Federal do Rio de Janeiro. Research productivity grantee by the National Council of Scientific and Technological Development (Conselho Nacional de Desenvolvimento Científico e Tecnológico - CNPq). Co-author of the Linguistic Atlas of Sergipe I (1987) and author of the Linguistic Atlas of Sergipe II (1995). Visiting Scholar at the Université Paris 13. Emeritus Editor of the Linguistic and Literary Studies. Member of the Brazilian Academy of Letters and Editor of the Academy Review. President of the Brazilian Association of Linguistics (1993–1995) and President-Director of the Linguístic Atlas of Brazil project (1996–2018). Deceased in 2018.

Jacyra Andrade Mota

Professor Emerita at the Universidade Federal da Bahia. PhD in Letters (Letras Vernáculas) from the Universidade Federal do Rio de Janeiro (2002) and Master in Letters and Linguistics by the Universidade Federal da Bahia (1980). Co-author of the book Livro das Aves (1965), of the Linguístic Atlas of Sergipe (1987) and of the Linguistic Atlas of Brazil. Vol. I e II (2014). President of the National Committee of the ALIB project. Professor in Residence of the Postdoctorate Programme in Language and Culture (PPGLinC). Research productivity grantee by the National Council of Scientific and Technological Development (Conselho Nacional de Desenvolvimento Científico e Tecnológico - CNPq). - Level I B. Research interests in Sociolinguistics and Dialectology.

Josiane Steluti is a Nutritionist, Master, PhD and Pós-Doc in Public Health by Public Health School - University of Sao Paulo (USP). Professor at Public Policies and Collective Health Department in the Federal University of Sao Paulo (UNIFESP) – Brazil. Collaborative member at Group of Epidemiology Studies and Innovation in Diet and Health/USP and Experimental Dietetics Lab - LaDEx/UNIFESP.

Dirce Maria Marchioni is a Nutritionist, Master and PhD in Public Health by Public Health School - University of Sao Paulo (USP). Professor at Nutrition Department in the Public Health School/USP - Brazil. Coordinator at Group of Epidemiology Studies and Innovation in Diet and Health (GEIAS)/USP and Group of Sustainable food Studies (Sustentarea)/USP and, member at Group of Planetary Health Studies – Institute of Advanced Studies/USP.

Charbel N. El-Hani is full professor in the Institute of Biology, Federal University of Bahia, Brazil. Coordinator of the History, Philosophy, and Biology Teaching Lab (LEFHBio) and the National Institute of Science and Technology in Interdisciplinary and Transdisciplinary Studies in Ecology and Evolution (INCT IN-TREE). Between January 2020 and July 2021, he was visiting researcher at the Centre for Social Studies, University of Coimbra, Portugal. He works in the areas of philosophy of biology, ecology, ethnobiology and science education research.

Rosiléia O. de Almeida is full professor in the Faculty of Education, Federal University of Bahia, Brazil. Member of the History, Philosophy, and Biology Teaching Lab (LEFHBio) and the National Institute of Science and Technology in Interdisciplinary and Transdisciplinary Studies in Ecology and Evolution (INCT IN-TREE). She works in the areas of environmental education, intercultural education, and science teaching.

Sávio Siqueira holds a PhD in Letters and Linguistics from Bahia Federal Univeristy (UFBA), Salvador, Brazil, and is an Associate Professor of English and Applied Linguistics at UFBA, has conducted Post-Doctoral studies at the University of Hawai'i Manoa, Honolulu, EUA. Permanent Professor at the Graduate Program in Language and Culture (UFBA) and the Graduate Program in Letters (UFT-Porto Nacional, Tocantins). He collaborates with the MA Program in Multilingualism, Education and Linguistics, Goldsmiths University of London and the Interinstitutional MA Program in Bilingual Education (MIEB), University of Jaén, Spain.

Acknowledgements

This book owes much to the work carried out within the scope of the following project (Sept. 2014–Dec. 2017)

"Glocal Languages" and "Intercultural Responsibility" in a post-colonial global academic world (GLOCADEMICS): *Power relations between languages/cultures within and between research groups* http://www.ces.uc.pt/projectos/glocademics

funded by an individual grant awarded by the European Commission (Marie Sklodowska-Curie Actions).

I am also most indebted to the Mary Sklodowska-Curie Actions Directorate because it allowed me to temporarily suspend this grant in order to assist my mother in her final dramatic months.

I am deeply grateful to the research groups' coordinators – Professors Suzana Cardoso and Jacyra Mota (*Universidade Federal da Bahia*), Professor Charbel El-Hani (*Universidade Federal da Bahia*), Professor Dirce Marchione (*Universidade de S. Paulo*), Professor Elizabeth Balbachevsky (*Universidade de S. Paulo*) and Professor Rosângela de Tugny (*Universidade Federal do Sul da Bahia*), as well as to the participating research group members, amongst whom I cannot go without awarding special notice, namely to the whole UFBA group of the ALiB project, Dr. Josiane Steluti (USP), Dr. José Verissimo (USP) and Professor Augustin de Tugny (UFSB). They all welcomed me and the Glocademics project with great professionalism and generosity.

I must also thank each one of the book contributors for accepting my challenge to add one chapter to this book and for their patience, support and perseverance throughout its long preparation.

I feel particularly thankful to the extraordinary personal and professional solidarity by the CES (Centro de Estudos Sociais) staff who backed me while I was away in Brazil and upon my return, namely its Executive Director, João Paulo Dias, the Project Manager, André Caiado, the Coordinator of the Project Management Section (GAGEP), Rita Pais, the ICT Technician, Pedro Abreu and the CES librarians, Maria José Carvalho, Acácio Machado and Inês Lima. I also thank Professor

Boaventura de Sousa Santos for accepting to be the overall coordinator of this project.

I thank *Fundação para a Ciência e Tecnologia*, Portugal (The Foundation for Science and Technology) for their support in attending some conferences where the content issues of the book were presented and discussed.

Last but not least, I am most indebted to Routledge staff and reviewers, for their careful readings and valuable recommendations, which allowed this publication to have materialized.

Introduction

Meta-reflections on critical and decolonial "glocademia"

Manuela Guilherme

The Inter- and Trans-nationalization of Higher Education and Research

Ours has recently been called "the fourth age of research" and is described as "international", having "progressed through three ages: the individual, the institutional and the national" (Adams 2013: 557). However, times have moved on quickly in this matter and globalization, internationalization of higher education, transnational funding policies and, of late, cost reduction in international scientific activities – aided by ICT facilities and the onset of national budget cuts – have resulted in increased staff mobility as well as job instability. Thus, it seems that we have rapidly arrived at a "transnational age", since what appears to be national is actually not and, consequently, is no longer inter-national. Unexpectedly, the COVID lockdown situation, and consequently the increase in online meetings and conferences, has enabled those with less funding for travelling, particularly from the Global South, to be able to participate in these events. They are therefore making their voices heard, both individually and collectively, that is, in their position of research group members, so as to present new research themes as well as different perspectives and innovative methodological approaches, although departing from scientifically established methods and criteria.

Yet, even previous to the COVID situation, the good news was that it seems possible to verify "the positive correlation between mobility and research quality" (Veugelers and Van Bouwel 2015: 363), in spite of the fact that trans-national research regulations, associated with tougher competition, may also result in contradictory consequences in terms of quality – over-intensive production *and* over-simplification of findings. Although Veugelers acknowledges the limitations of her data due to language and discipline constraints "as recorded by Thomson's ISI-Web of Science journals", being both monolingual – in English – and limited to each particular discipline, she still finds evidence of a trend toward "a slow but real process of increasing convergence, with the catching-up of non-TRIAD countries"[1] (2010: 441–448). According to her,

DOI: 10.4324/9781003225812-1

this finding justifies the following question: "Does the increasing rise of non-traditional science countries manifest itself in changing patterns of international collaboration?" (ibid.). With a clear emphasis on research generated in Latin America among the new contributors, most of their relevant and innovative studies in the social sciences being published in Portuguese or Spanish, these naturally stand some distance apart from Veugelers' study.

The contributors to this book, myself included, agree that, in response to Veugelers' question, we should tend to open up patterns, standards and methodological approaches, without bringing rigor, validity or credibility into question. For that purpose, we must trigger new partnerships and co-operations, for mutual learning and local-global findings that allow inter-epistemological exchange and, therefore, meet local/global demands, improve the life conditions of target populations, generate epistemological equity and, ultimately, social, and cognitive justice (Santos 2018). Both flooding or draught and fires, originated in climate change, and the COVID pandemic have put in alarming evidence the need for an inclusive transnational, inter-epistemologically decolonial, intercultural, and plurilingual academy that promotes an "ecology of knowledges" (Santos 2014). In sum, there is a pressing need to critically re-think the "glocal" – simultaneously and, as much as possible, symmetrically global and local – role of academics and researchers, hereafter referred to as "glocademics", and, thus, their performance of linguistically and culturally aware research procedures, both within the context of research and education. To be more precise, as the impact of the global in the local has earned the focus from the studies on the internationalization of higher education, it is our goal here to highlight the potential of bottom-up transnational contributions by local research projects, nonetheless globally informed, both targeted at local environments and enriched as well as broadened, through reciprocal import/export exchanges, by transnational collaboration.

The Internationalisation of Higher Education Institutions (IoHEIs) has generated much scholarly interest since the turn of the century, mainly concerned with policy decisions about the knowledge economy, such as the European Higher Education Area which has also inspired higher education planning in other regions of the world, including Latin America (Tiana-Ferrer 2014). The scholarly interest in the field has gained momentum worldwide and produced a bibliographic *corpus* sufficiently relevant to be considered as a specific field of expertise (Miuht et al. 2017). The literature on IoHEIs has nevertheless focused on the ways in which HEIs have had to adapt to global pressures locally while simultaneously attending to local needs (Altbach 2004). The same has happened with research which, in order to transcend national borders and become transnational, has had to comply with global criteria, for example, the English language usage with respect to academic register.

Altbach (2013) and De Wit (2011) have introduced these issues of monolingualism in English and intercultural competence into the discussion on IoHEIs. As such they have acknowledged that these aspects have been neglected in this regard but they have failed to dig deeper into them. To sum up, they have concluded that greater involvement at international or transnational levels, does not necessarily result in greater intercultural emphasis. However, another decade has elapsed and funding institutions, specifically those linked with the European Commission, such as the European Research Council, or the US-based Spencer Foundation, just to name a few, have not funded research with such target.

This claim will present important challenges in the near future to the "inter-continentalization" of the knowledge economy, a goal long established by the European Commission and by all other regions in the world as well. All in all, it has become evident that globality in decolonial times requires that each and every researcher reappraises their position, from their own locality, and that the spheres where knowledge is created become an imaginative and prolific environment where mutual learning can prosper (Banerjee *et al.* 2016). In India, scholars are also laying claim to a re-positioning not only of residual internal colonialism but also of post-colonialism (Spivak 1999, Visvanathan 2007). Scholars from Europe and North America have also pointed out new paths for building on the concept of an international/intercontinental university. Santos, for example, argues that Europe needs to engage in a dialogical and horizontal positioning with the rest of the world mainly with respect to epistemological diversity for which he proposes the encouragement of an "ecology of knowledges" (Santos 2014). Furthermore, academics in the Global South are also vindicating their epistemological parity and the possibility to make the credibility of their research outcomes recognized by the global community, as below. This debate is closely connected with the contemporary discussions on post-colonialism and decoloniality led by scholars in the Global South, as I shall briefly attempt to outline below.

The Challenges of the South-North Relationship in Decoloniality

We can identify some historical, political as well as theoretical milestones in recent history which have established some steps in the self-representation of the idea of the South. To start with, the Bandung conference in 1955, which was held in Indonesia, aimed to discuss the role of the Third World post-colonial countries, namely in South Asia and Africa, in the early Cold War times and their position in relation to the two opposing Northern hemisphere blocks, Western Europe and the United States, united in a post-Second World War alliance, on the one hand, and the Soviet Union, on the other hand. In the Opening Speech,

President Sukarno proclaimed that they, the leaders of the recently independent countries, had earned "heavy responsibilities to [themselves], and to the world, and to the yet unborn generations" and warned that: Colonialism [had] also its modern dress, in the form of economic control, intellectual control, actual physical control by a small but alien community within a nation. It [was] a skilful and determined enemy, and it appear[ed] in many guises. It [did] not give up its loot easily" (https://www.southcentre.int/question/revisiting-the-1955-bandung-asian-african-conference-and-its-legacy/). President Nehru, from India, and President Nasser, from Egypt, were also present among the twenty-eight leaders attending the conference. The Popular Republic of China, a third emerging power, represented by Chou En-Lai, also played a leading role in the Bandung Conference while consolidating its double-edged position vis-à-vis the two main blocks.

A second historical and political milestone was the Report of The South Commission, Oxford University Press (1990), https://www.southcentre.int/wp-content/uploads/2013/02/The-Challenge-to-the-South_EN.pdf), chaired by President Nyerere from Tanzania. Both the Bandung Conference Opening Speech and The Challenge to the South Report underline the importance of "responsibility" taken by the South. The Report concludes by reminding its readers that "The South covers the larger part of the Earth's surface. Its people are the vast majority of the world's inhabitants" and by listing 'the challenges to the South' of which the final statement is that "The responsibility to work for change in the present conditions therefore lies firmly with the South" (p. 24). The use of the term 'responsibility', hereby used, brings in a new concept, that of an emancipatory idea of control of one's life by the disadvantaged, rather than a patronizing and humanitarian idea of 'responsibility' of the stronger towards the weaker, instead one perceived as a right imbedded in reciprocal and respectful partnership. The idea of "responsibility" as "duty", in the perception of post-World War II Jewish thinkers, such as Levinas (1998), *versus* bottom-up "right", in the sense of power earned through a 'decolonial turn' (Castro-Gómez and Grosfoguel 2007).

There is recent, vast, and relevant literature on post-colonialism and decoloniality which would be extremely valuable and of utmost pertinence for a contextualization of the challenges addressed by the various chapters of this book, however, due to word limit constraints, I can only provide a very succinct selection. Therefore, I will restrict this brief introduction also to some theoretical milestones which can illustrate different approaches and the progressive development of this perspective. Bhambra (2014) examined the differences between postcolonial and decolonial dialogues, mainly in terms of time, the former setting its roots mainly in the nineteenth and twentieth century, and the latter developing in contemporaneity but also drawing inspiration from pre-colonial traditions and civilizations. Bhambra (ibid.) rightly points out the importance of space in determining the vision towards this process of building and validating Southern identities and voices, either as postcolonialism or decoloniality, although most of the scholars establishing the foundations for both approaches have been in diaspora in North/West universities such as E. Said, H. Bhabha, G. C. Spivak, W. C. Mignolo, R. Grosfoguel, etc. However, it is also important to highlight the fact that postcolonial and decolonial studies have been developing exponentially in research centeres and higher education institutions located in the Global South, including permanently established diasporic scholars from the Global North as well, for example, C. Walsh at the Universidad Andina Simón Bolívar in Quito, Ecuador. In addition, we cannot go without pointing out how the conceptual frameworks and use of both terms, **post**-colonial**ism**, likewise to **multi**cultural**ism**, and decolonial**ity**, which is foundational to (critical) **inter**cultural**ity**, also depend on the colonial matrix from which they stem (Guilherme and Dietz 2015, Guilherme 2019a, Guilherme forthcoming).

Spivak (1988) raises crucial aspects in her critique of (post)colonialism which cannot be overlooked either by post- or de-colonial transnational researchers wherever they are placed or how they are ideologically positioned. In her seminal text "Can the Subaltern Speak?", she questions the notion of the Other, which after all intends to keep "the West as Subject" (p. 66), and whom she calls the "unnamed subject of the Other of Europe" by describing 'it' as the "Self's shadow" (p. 280) and whom she still considers a "colonial subject" (p. 281). Furthermore, she reminds her readers that "one must nevertheless insist that the colonized subaltern *subject* is irretrievably heterogeneous" (p. 284) and that its liberation will emerge from their own assumption of themselves, both individually and collectively. Later, Spivak (2004) described the "subalterns" as "those removed from lines of social mobility" and, although she is critical of much of European and North American scholarly writing on power, it is along these lines that she carries out a sharp appreciation of the concept of responsibility which has, according to her, been progressively imbued with a sense of duty, the "'duty of the fitter self'" (p. 535), and less perceived as a right. She is clear that this is often the point of view of dominant transnational organizations, such as the UN, and of leading politicians of the North and the West, such as Churchill, as she identifies a "Churchillian sense of 'responsibility'", while she adds that "the question of responsibility in subordinate cultures, is also a begged question" (p. 535). In sum, Spivak interrogates above about several subtleties which, in general, remain unquestioned in debates about postcolonialism.

Nevertheless, both theories on postcolonialism, which in the seventies had already spread worldwide, as well as Eurocentric pluralist post-World War and postmodern critique, inspired and provided the background for the emergence of decolonial studies in Latin America which have nevertheless taken off from those grounds. However, there is no conflicting positions between one and the other but a rupture of both, albeit in different moments, spaces and colonial traditions, with the "coloniality of power", the "coloniality of being", and the "coloniality of knowing". That is, the scholarship about decoloniality that develops in Latin America, formally since the 1970s, but originally since the European invasions (discoveries), as it is claimed, proposes a "decolonial turn" which captures, validates and re-creates the inner soul, the heritage and the indigeneity of Latin America or whichever postcolonial location (Castro-Gomes, Dussell, Grosfoquel, Mignolo, Torres, Walsh, etc., as we shall see below). According to Mignolo, the Philosophy of Liberation proposed by Enrique Dussel (see chapter 8 later in this book), "comes from the subaltern perspective – not from the colonial/Christian discourse of Spanish colonialism but from the perspective of its consequences, that is, the repression of American Indians, African slavery and the emergence of a Creole consciousness ..." (2008, p. 234). In addition, Dussel (2008) stated that the Philosophy of Liberation, closely connected with the idea of decoloniality, was not exclusive to Latin American thought or history but a "critical philosophy self-critically localized in the periphery" (p. 340), which he admitted to have initiated simultaneously to the foundation of the "subaltern studies" in India which were later enriched by Spivak's work, as mentioned above. And, in Mignolo's words (2000, 2007, 2008), decoloniality encompasses the idea of 'border thinking' which has also been included by the theoretical background on post colonialism, although the term was originated within the scope of decolonial studies, first used by Gloria Anzaldúa in her book "Borderlands/La Frontera: The New Mestiza".

From the perspective of decolonial studies in Latin America, the "decoloniality of knowledge" encompasses the acknowledgement of the *locus/loci of enunciation*, which are not necessarily singular, homogeneous, stable or self-sufficient entities. Instead, decoloniality dismantles what Castro-Gómez (2005) calls "La Hybris del Punto Cero" (The Hubris of the Zero Point, my translation), a God-like point of view from where you can examine everyone else but no one can examine you, neglecting reciprocal dialogue and "intercultural translation" (Santos 2018) and certainly inter-epistemic discussion. On "the epistemic decolonial turn", Grosfoguel (2007) made clear that he was not claiming for a kind of "epistemic populism", instead decolonial thinkers are simply demanding that no source of knowledge can be entitled as such, "The Hubris of the Zero Point".

Mignolo has been concerned with what he calls "the geopolitics of knowledge" and has introduced several concepts in this regard, such as "epistemic de-linking" which is all about disconnecting from the colonial matrix of power/being/knowledge and by which he means that "both 'liberation' and 'decolonization' points towards conceptual (and therefore epistemic) projects of *de-linking* from the colonial matrix of power" (2007, p. 455). He also calls this stand as "epistemic disobedience" which concerns not only decolonizing knowledge but also the process of knowledge-making (Mignolo 2009), that is, "delinking means to change the terms and not just the contents of the conversation" (2007, p. 459).

Mignolo (2018) makes the distinction "between decolonization during the Cold War and decoloniality after the end of the Cold War" and he adds that "coloniality- the darker side of Western modernity – is a decolonial concept and therefore the anchor of decolonial thinking and doing in the praxis of living" (pp. 106–107). These are important conceptual clarifications for those who deal with 'knowledge-making' in any research community of both academics and non-academics (the latter can be temporary community researchers in academic projects) which are subject to discussion as well. For example, in some Critical Dialogues in response to Mignolo's and Walsh's book on Decoloniality (2018), Gu (2020) from China, adds that "decoloniality should be concerned with the present condition shaped by the colonial past" (p. 598) while Ndlovu (2020) from South Africa, raises "a number of questions around the effectiveness of decoloniality as a 'combative tool'" and warns that "though it produces a paradigmatic shift from the totalizing and universalist approach to knowledge and life, it also opens up decoloniality for misuse and abuse" (p. 580). However, the scholarship on decolonial studies is far from being a monolithic block and it is subject to internal critique. Cusicanqui (2019) from Bolivia, for example, also warns about "new forms of colonization" and "a new academic canon" which can be concealed in the mechanisms of "official multiculturalism" and the discourses of "multiculturalism and hybridity" while Makoni (2019) concludes that "decoloniality is still however, a very contentious strategy because the term means different things to different people" (p. 150).

A decolonial concept of "critical interculturality", with a perspective from the South and based on indigenous cosmologies in the west, north and center, of Latin America, has been developed by Catherine Walsh and become a fundamental resource for the development of this concept underlying the European-Latin American projects which link the various chapters in this book. The element of critical interculturality is inherent to decoloniality and vice-versa, one cannot exist without the other, and Walsh captured this interrelation in depth (2012, 2018). Her work can be considered an example of "lived" decolonial theory developed through her search as an activist and intellectual (I would say as a Glocademic, described below) for the spirit of the South of Abya Yala (the Americas) whose internal power struggles cannot be ignored either. Violence *tout court*, violence on women, on the elderly and on children, as well as racist discrimination, run across all geographies, not in the same degree of legality, but they are there nonetheless. Besides, ethnocentrism is the best distributed thing in the world, in the words of the Brazilian anthropologist Viveiros de Castro (Castro 2015).

Dissecting the concepts of "critical" and "intercultural" from a decolonial perspective, and from the perspective of a metaphorical South, as explained below, helps building a decolonial, plurilingual and glocal academy of researchers in their commitments towards knowledge creation and civic commitments. Walsh's work (e.g. 2012) has been seminal in this trajectory towards finding other philosophies and other knowledges that may challenge the tight epistemological boundaries which have been established by a tiny circle of knowledge workers. Walsh highlights "the importance of 'other' critical modes of knowledge ... that enable critical theorizations *from difference*, opening up new analytic, critical, post/trans-continental, and decolonial possibilities of knowledge and existence" (2012, p. 15). A kind of functional consensual interculturality acting at the surface of diversity management cannot support any critical, decolonial, and reflexive knowledge-making project, even less those which are based on a "research-on-research" process as the ones described in this book. Critical interculturality, from a decolonial perspective, departs "from the particularity of local histories, and political, ethical, and epistemic places of enunciation, all of which are marked by the colonial difference and by decolonial struggle ... [and it] extends its project of an *otherwise*, a transformation conceived and impelled from the margins, from the ground up, and for society at large" (Walsh 2018, p. 61).

In the same lines of critical and decolonial thought, in the introduction to his book on "The Epistemologies of the South", Santos (2014) is clear about his overall argument: "... it is imperative to go South and learn from the South, though not from the imperial South (which reproduces in the South the logic of the North taken as universal) but rather from the anti-imperial South", which means that it is also "imperative to start an intercultural dialogue and translation among different critical knowledges and practices" (p. 42). For such a purpose, Santos offers some leading ideas, which run throughout his work and support his umbrella theme of an "ecology of knowledges". Among those, it is relevant here to point out "intercultural translation", "post-abyssal thinking" and "sociology of emergences", all of which support a praxis that not only makes evident the reciprocal incompleteness, interdependence, and complementarity of different kinds of knowledge but also enrich the possible approaches of research and education with regard to the parameters defining knowledge and to the agents of knowledge creation. The expressions above stand out for different conceptual strategies which are indispensable for a new administration of knowledge creation, credibility and sustainability, and therefore inspiring for the approach taken by the research experiments described in the chapters that follow. Respectively, 'intercultural translation' which is described as "a living process of complex interactions among heterogeneous artifacts, both linguistic and nonlinguistic, combined with exchanges that by far exceeded logocentric or discourse-centric frameworks" (Santos 2014, p. 215), promotes the critical intercultural awareness of different conceptual frameworks which nevertheless allows dialogue, comparability, and exchange.

However, "intercultural translation", within the scope of an "ecology of knowledges", also keeps in mind that the terms of conversation between different types of knowledge are not yet entirely reciprocal because the power relations between them are still asymmetrical, and require that a kind of "post-abyssal thinking" is developed and puts down the frontiers between pre-determined epistemological zones of light and shadow (Santos 2007). It is in this space where knowledges, those made visible as well as those made invisible through processes of coloniality, were simultaneously built, in a process which Santos (2018) describes as a kind of Janus-faced sociology, composed of both "absences" and "emergences". And it is in the same space that the abyssal line established in-between is confronted and provides the foundations for the constitution of the Epistemologies of the South. While the "sociology of absences" has lingered at the macro-level, the reconfiguration of a "sociology of emergences" demands that such movements are articulated between the micro- and macro-levels, which puts the focus on what had been made invisible but "has to be transscale" (Santos 2018, pp. 250–251). Eventually, knowledge creation and application needs to be contextualized both at the macro- and micro-levels. According to Santos, scientific knowledge has progressively tended to become decontextualized, in order to grow more technical, however, in his words "once decontextualized, knowledge becomes potentially absolute", even more so whenever knowledge which is not technical has been simply neglected (2006, p. 31)

The "decoloniality of knowledge", which is irredeemably tied to the "decoloniality of power" and the "decoloniality of being", is at the core of this book. Dussel, an Argentinian exiled in Mexico, affirms that "the philosopher of liberation neither represents anybody nor speaks on behalf of the others ... that is, it assumes responsibility for all sorts of alterity. And it does so with an ethical, 'situated' consciousness ..." (2008, p. 342). Colombian Fals-Borda and Mora-Osejo (2007) also supports such research ethos within the vision of a Philosophy of Liberation by stating that: "Scientific insight and authority come from this involvement with real life" (p. 401). Although this vision of research does not preclude inter- or trans-national collaboration, it calls for non-extractivist methodologies which Santos describes as "postabyssal scientific knowledge" generating "coknowledge emerging from processes of knowing-with rather than knowing-about" (2018, p. 147). This book aims to share a process of decolonial "co-learning" which results from intensive critical and metareflexive dialogue **with** and **between** "Glocademic" researchers (see below) who implement that practice in their research routines and who accepted to offer, in direct speech, their critical reports and analyses of their experiments and experience.

It was in the aftermath of the ALFA-RIAIPE 3 project (https://www.ceied.ulusofona.pt/en/directory-research/riaipe3/), and simultaneous to the ERC-ALICE project (https://alice.ces.uc.pt/en/?lang=en), that the MSC-Glocademics matrix (https://www.ces.uc.pt/ces/projectos/glocademics/) was put together and developed. It also built upon research work carried out through Leonardo da Vinci-ICOPROMO (https://www.ces.uc.pt/ces/icopromo/) and Framework VI-INTERACT projects (https://www.ces.uc.pt/ces/interact/). The projects above concentrated their attention in the Epistemologies of the South, from the point of view of critical interculturality, and the South-North inter-epistemological relations, and were located in different geographies, mainly in Latin America and Europe.

The Role of Glocademics in Critical Transnational Research

The development and institutional validation of transnational research requires considerable meta-reflexive attention regarding a new professional identity, such as that contained in the "glocademics" concept, meaning those who work at the grassroots level in order to build a responsible and socially encompassing glocal academy, responding to local and regional needs while also benefitting from transnational exchange and collaboration. Glocademics need opportunities to engage in interdisciplinary "research on research" (Enserink 2018) projects and specific development programs, both of which may help them operationalize their plurilingual and critical intercultural awareness in order to reach rigorous and scientifically valid inter-epistemic knowledge.

This edited book to go beyond a single-author vision, one which provides an individual take on plurilingual, intercultural, and inter-epistemic "research-on-research" experience. Hence, it aims to take a decolonial approach to research carried out in the geographical South by mainly giving the floor to academics who have developed it in praxis, both those of European origin, namely Portuguese, with experience in Europe-Latin America collaboration projects, who contribute with two introductory projects, and a greater part of Brazilian academics who take a critical decolonial approach to South-based research. The book editor and project principal investigator aims to wrap up their message by taking advantage of a South-North look she has developed throughout her career and life experience. Through the implementation of the Glocademics project: *'Glocal Languages' and 'Intercultural Responsibility' in a postcolonial global academic world: Power relations between languages/cultures within and between research groups*, the book editor/principal investigator invited her peers to engage themselves in a meta-reflexive process concerning their own research projects from the point of view of linguistic, cultural, and epistemic diversity, the specific challenge launched by the Glocademics project.

The research landscape has become "glocal", both global and local, and therefore transnational. National identity and citizenship have remained meaningful, while regional and local contexts are still significant but with the broader context of knowledge exchange and intercultural comparability now offering limitless potential. So as not to become meaningless, the focus cannot be blurred. Rather there is a need to remain fully concentrated on the details, complexities and entanglements emerging in-between the global-local dynamics. Hence, a comprehensive conceptual framework was found necessary, one that is enlightened by already existing bodies of knowledge which aim to respond to current challenges in knowledge creation and research management and, at the same time, proposes new paths for approaching the plurilingual, intercultural and inter-epistemic dimensions in that venture.

The GLOCADEMIA theoretical framework draws on three interdisciplinary, theoretical and practical bodies of knowledge which have undergone intensive development in the last few decades, namely the internationalization of higher education, which has, to some extent, addressed issues related to academic mobility, internationalization of academic institutions, but which require further development. Great significance is also attached to the field of intercultural communication and education, which has produced a vast bibliographic corpus. Although it has tended to take on an essentially functional approach, ranging from that of social pragmatist when faced with neoliberalism and neo-colonialism – on the grounds of improving commercial relations – through to youth employability and mobility, most of its work has also included calls for individual reflexivity and mindfulness, more often within communication and psychology studies, and/or collective awareness of cultural diversity, intercultural dialogue, and intercultural citizenship, in general terms, promoted through education and sociological studies. Finally, interdisciplinarity has raised significant scientific interest from experts in different disciplines as well as policymakers, both for cross-disciplinary fieldwork and that operating within the same discipline, and is vital to stand out as a third element within the umbrella themes.

All three bodies of knowledge have opened up new relevant paths which enable us to carry out 'research on research' on trans- and intranational research activities. The three umbrella themes in Figure 0.1 illuminate a three-pillar- based conceptual framework, namely: (1) Glocademics; (2) Glocal Languages; and (3) Intercultural Responsibility, all of them geared to addressing the possibilities and tensions of global-local dynamics in research activities. The 'glocal' backdrop departs from

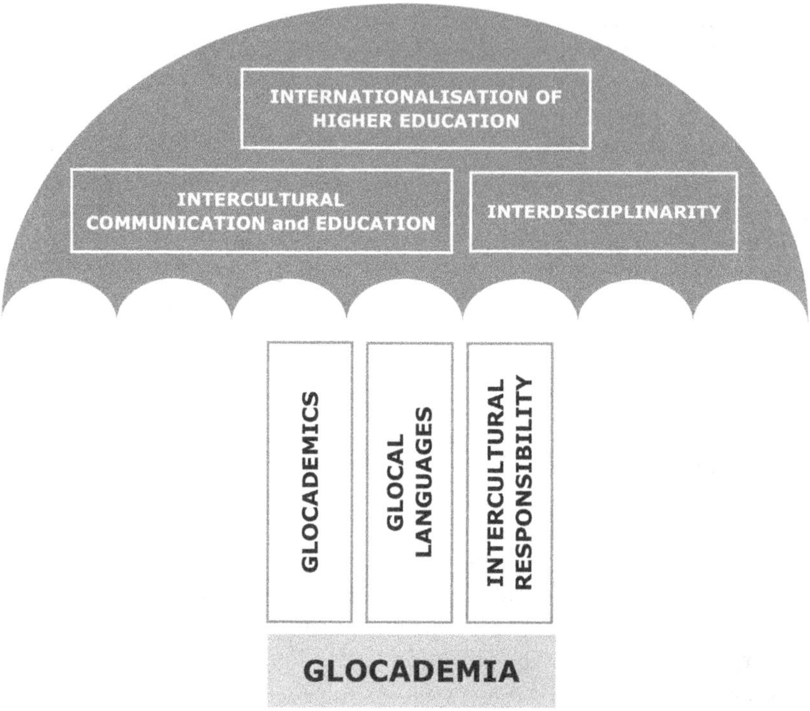

Figure 0.1 GLOCADEMIA theoretical framework

the idea of "glocalization" put forward by Robertson (1995) and from Santos' notions of "localism" and "globalism", a critique of the forms of production of hegemonic globalization highlighting the process through which one localism becomes globalized and then localized elsewhere (1999). However, the term "glocal", within the scope of the glocademia conceptual framework relies on a decolonial approach through which a globalism, at once localized, can incorporate the capacity to "answer back" (Guilherme 2019b) and "think otherwise" (Mignolo and Walsh 2018). The "Intercultural Responsibility" pillar will be further discussed below (Guilherme 2020a/b). The following chapters aim to illustrate the Glocademia experience and make visible the inner workings of its three axes shown above in Figure 0.1 and which guided the research questions of the Glocademics project (https://www.ces.uc.pt/ces/projectos/glocademics/).

The main goal of the EC-MSC (European Commission – Marie Skowdlovska- Curie) Glocademics project was to open up a new research path, that of meta-reflexive "research on research", while aspiring to be ground-breaking in epistemological terms. Yet, this book is not restricted to an individual account of the Glocademics project. Rather, its aim is broader in providing access to an emblematic illustration of the glocademia world through the meta-reflexive we-voices of glocademics who are providing new ground in their own research activities and knowledge-producing spaces, in their condition of glocal sites of citizenship, through transnationalization and relocalization of academic global and local life. The reader can find below meta-reflexive and critical descriptive and narrative accounts of research projects in the life sciences and the social sciences as well as the humanities, the majority of them carried out mainly in Brazil by researchers from different regions. Those involved in the projects represented in this book were implementing plurilingual and intercultural research activities, some of them dealing with knowledge going beyond a simplistic concept of science, yet knowledge still translatable into scientific analysis, translated from various languages with a dominance of Portuguese, Spanish, and English. This book offers, in unison with the aims of the Glocademics project, the opportunity for a meta-reflection on the glocademia experience, that is, an interweaving of the global and local principles, interests and needs, both from an individual and collective point of view.

The title of the book also deals with specific prefixes added to certain adjectives, namely **pluri**lingual, **inter**cultural, **inter**epistemic and **trans**national, which I shall attempt to justify albeit briefly. The reasons for the choice of these specific prefixes are not random. According to the Common European Framework of Reference for Languages (Council of Europe 2001), "plurilingualism" refers to the individual use of several languages (pp. 4–5). In this case, "plurilingualism" refers not only to the individual researcher but also to the linguistic baggage of another singular entity, the research group. The term "intercultural", alongside general usage of the prefix inter-, which has become more common in every context over the past two decades also admits a wealth of hermeneutical possibilities, although the whole title and goal of the book only embrace the notion of critical interculturality which is discussed throughout the book and encompasses the significance of "inter-epistemic" collaboration (Guilherme and Dietz 2015, Guilherme 2019a). The scope of the 'transnational' horizon in research collaboration does not neglect the researchers' national loyalties or their local intra-national roots, although multiple, mixed and complex they may be, and consequently the idiosyncrasies of knowledge collection, exchange and recreation across languages, cultures and epistemologies. Yet, all of these terms are endowed with multiple meanings and implications, embedded in colonial matrices which cannot be disregarded when critical and decolonial approaches to cultural diversity are undertaken, which is the case here.

The first two chapters offer an introduction to the general theme of the book and to the work presented in the remaining chapters which stand more closely related to the Glocademics project mentioned above. While chapter 1 interrogates about core issues regarding postgraduate programs, whose structure and goals were originally exported from

Europe, is still dominant in universities around the world and remain deep-seated in the academic landscape described in the following chapters, chapter 2 gives voice to Latin American academics participating in a large European funded project. Chapter 1 introduces this book not only because it is assumed that international and inter-institutional doctoral programs are breeding grounds for future transnational researchers, while questioning the aims of doctoral education with regard to research and society, but also because it focuses on EHEA policies and their political assumptions, which have been inspirational worldwide. The authors pertinently bring to the fore a critical view of the aims of the "knowledge society", its goals relating to innovation and how it impacts on doctoral programs, that is, on the formation of research workers, ultimately transnational researchers-to-be, those referred to here as "glocademics". The authors examine the contradictions between the diversity of students, not necessarily European, integrating doctoral programs which are nevertheless awarded with institutional autonomy, as opposed to the hegemonic model of monocultural and mono-epistemological knowledge creation which, ultimately, intends and pretends to respond to the alleged needs of the labor market and one which is expected to translate into the epitome of the 'entrepreneurial' university. In sum, the authors put their focus on a model which is nonetheless germane to this book's argument given that this political and epistemological framework is transnationally, and often uncritically, being embraced worldwide.

Chapter 2 then introduces some elements on European - Latin America research cooperation which allows the organization of trans-national research networks with a focus on the cooperation between European and Latin American institutions and research teams. It starts by describing the foundation of the Ibero-American Research Network in Education Policies (RIAIPE) and its expansion into the Inter-University Framework Program for Equity and Social Cohesion Policies in Higher Education (RIAIPE3) project. Last but not the least, the chapter gives voice to a large number of the researchers involved in this project, both European and Latin American, who video-recorded their views on the functioning of this large three-year project network during the final meeting at the *Universidade Nove de Julho* (UNINOVE) in S. Paulo (2013). Not only do they reflect on the project activities but, more specifically, on the terms of collaboration among the project participants, more particularly between European and Latin American researchers. Such recordings, which can still be found online (see chapter references), provide a unique video testimony of 26 senior researchers representing 25 universities (19 from thirteen Latin-American countries and 6 from four European countries).

Chapter 3 proceeds with an analysis of the democratic developments in Brazil in recent times through an international comparative study between Brazil and England based on conceptual analysis from the viewpoint of political scientists. The text shares a critical reflection on both democratic institutional designs and their role in the quality of democracy by focusing on emerging differences of newly implemented forms of institutional "flexibility" and "incompleteness". This project was selected as one of the focal points of the Glocademics project due to the interesting cross-cultural, cross-linguistic and cross-national conceptual analysis it was undertaking. The chapter offers first-hand perspectives and voices of two of its participants and their meta-reflections on the comparative study of common concepts which both research groups, in Brazil and in England, had selected to target. The authors conclude by demonstrating how crucial intercultural conceptual analysis has become for inter- or trans-national research projects and, therefore, the importance of critical intercultural awareness in this process. To sum up, this chapter emphasizes the relevance of comparative studies regarding different ways of institutional functioning for the purpose of evaluating implementation of public policies and their effectiveness.

Chapter 4 provides the meta-reflection of another Glocademics project's focal research groupwork undertaken by their coordinators. It provides us with a hint of the enormous amount of data gathered and analyzed by The Linguistic Atlas of Brazil (ALiB) research group over the last few years. The project is a milestone in Brazilian linguistics given its nationwide coverage and its capacity to remain intact for many decades through individual grants only, serving to underline its significance among the linguistic world atlases due to the dimensions of the territory covered. The authors give an overview of the project's widescale scope which has provided enough data both for several sub-projects with regional and thematic coverage and for transnational cooperation, alongside the main bulk of the project's publications which already extends to several volumes. The chapter illustrates the possibilities of undertaking cross-national comparative studies through a mix of national and inter-national foci. According to the authors, the aim of the project was to unearth all the idiosyncratic features of contemporary Brazilian Portuguese, namely diatopic, diastratic, and diagenerational variation, differences in pronunciation, semantics, etc., of which the authors provide substantial evidence, notwithstanding the difficulties of translating them into English. The text also unveils the remains of a colonial matrix which relied greatly on miscegenation. Besides manifold examples of older Portuguese which are uncommon in contemporary versions of European Portuguese, there exist instances of immigrant and indigenous languages, the latter displaying a remarkable presence in Brazilian toponymy. The focus of the ALiB project is the Brazilian Portuguese language in all its varieties and influences. Given the extent of this focus, it is unable to include - though it does not discard - other languages that remain in Brazil. The authors also give examples of the input of immigrant languages, either resulting from groups of immigrants arriving on a voluntary basis such as Italians, Germans, Japanese, Lebanese, etc. and from involuntary groups of immigrants, namely African slaves whose influence in gastronomy and dance in the northeast is noteworthy. This large group of researchers have been searching for the historical layers of linguistic colonization and working hard, in association with local populations, to make visible the re-creation of world epistemologies inside Brazil through the Brazilian Portuguese language.

Chapter 5 describes the work carried out by the Brazilian research team whose members represent different universities in various regions of the country when translating, adapting, creating, and testing a Brazilian version of the GloboDiet software previously designed and implemented in European countries. This software was originally designed in English for a project coordinated by the International Agency for Research in Cancer (IARC) whose headquarters are based in France. The authors narrate the history of this project, its different steps and illustrate the accurate linguistic and (inter)cultural 'translation' of the software into the rich semantic, agricultural and gastronomical diversity of the multiple and largescale geographical and cultural regions of Brazil. This software was designed with the intention of carrying out national surveys aimed at gathering standardized dietary intake data into common global and regional databases for future research and which may be used to support public health policies. For the purpose of the Glocademics project it offered a wealth of data to be analyzed through the lenses of critical intercultural translation.

Chapter 6 accounts for another research group serving as a focal point for the Glocademics project. The chapter offers the authors' own critical reflections upon university-community partnerships and educational proposals for intercultural dialogue between school knowledge and traditional knowledge. In this case, their research work has concentrated on the bodies of knowledge of traditional fishing communities in the northeast of Bahia. The authors introduce a model of inter- and trans-disciplinary research and innovation in traditional communities through a participatory process aimed at empowering the community by including their goals and by respecting their own understanding of how to achieve better life standards. Along the way, they have managed to build bridges between the natural sciences and the humanities.

Chapter 7 consists of an interview by Sávio Siqueira, from the *Universidade Federal da Bahia*, to the book editor, in her position as the author of the proposal approved by the European Commission and principal investigator of the Glocademics project. The interviewer, who closely assisted the implementation of the project at UFBA, elicits a personal account of the project at UFBA with the intention of illustrating the accomplishing of the project at other sites alongside this example. The implementation of the Glocademics project at UFBA was chosen for this concluding interview because, due to the design of the university campus and the fact that both participating research groups were settled in buildings close to each other, it was possible to organize a so-called "interdisciplinary event". More than a dozen participants, of a larger group, in the Glocademics project kindly agreed to sit face-to-face and discuss for a few hours the issues raised by the project concerning national and transnational research. The contents of this interview very much express the reflections brought up by this get-together at UFBA as much as the interviewees' own reflections upon the results of this project. Moreover, Dr. Siqueira had played the role of a distant but well-informed observer of the development of the project at UFBA.

Finally, Chapter 8 contextualizes the concept of 'intercultural responsibility', the third axis of the Glocademia matrix, within the broader attempts of transnational organizations to link the idea of responsibility with research planning and implementation. The chapter starts by introducing the discussion about the ideas of Responsible Research and Innovation (RRI) and Responsible Innovation (RI), which have been promoted by the European Commission research programs, with specific reference to their reception in Brazil. It goes on to examine the tradition of university extension, one of the three main pillars of university life in Latin America that are meant to unfold together, and which may add to the idea of responsible research. It then proposes that the theme "science in society" be altered to "society in science". This chapter also undertakes an analysis of some theories on responsibility put forward after the Second World War by Jewish authors such as Jonas, who presented demands for a new ethics that responds to technology and the vulnerability of nature, Arendt, Levinas, and others, who paved the way for a pluralist notion of responsibility. The chapter aims to move beyond the limits of a Eurocentric understanding of the notion of responsibility and suggests that an intercultural perspective of responsibility can also be expanded towards the Latin American proposal for a "philosophy of liberation", propounded by Dussel and others, and supported by the "epistemologies of the south", proposed by Santos, both of which also contribute to a "decolonial turn", proposed by Mignolo, Castro-Gomez, Walsh, etc. This chapter seeks practical support following the results of the Glocademics project, mentioned earlier, in order to clarify the implications for research which may be brought about by this third pillar of the glocademia matrix, that of intercultural responsibility.

Note

1. TRIAD countries – North America, Western Europe and Japan (https://ec.europa.eu/eurostat/statistics-explained/index.php/Glossary:Triad)

References

Adams, J. (2013). The fourth age of research. *Nature*, 497: 557–560

Altbach, P. G. (2004). Globalisation and the university: Myths and realities in an unequal world. *Tertiary Education and Management*, 10:1, 3–25

Altbach, P. G. (2013). *The International Imperative on Higher Education.* Rotterdam: Sense

Banerjee, P., Nigam, A. and Paney, R. (2016). The work of theory: Thinking across traditions. *Economic and Political Weekly*, 51:37, 42–50

Bhambra, K. G. (2014). Postcolonial and decolonial dialogues. *Postcolonial Studies*, 17:2, 115–121

Castro, E. V. de (2015). *Metafísicas Canibais.* São Paulo: Cosac-Naify

Castro-Gómez, S. (2005). *La Hybris del Punto Cero: Ciencia, raza e ilustración en la nueva Granada (1750-1816).* Bogota: Pontificia Universidad Javeriana

Castro-Gómez, S. and Grosfoguel, R. (2007). *El Giro Decolonial: Reflexiones para una diversidade epistémica más allá del capitalismo global.* Bogotá: Siglo del Hombre Editores

Council of Europe (2001) *Common European Framework of Reference for Languages: Learning, Teaching, Assessment.* Cambridge: Cambridge University Press

Cusicanqui, S. R. (2019). Ch'ixinakax utxiwa: A reflection on the practices and discourses of decolonization. *Language, Culture and Society*, 1:1, 106–119

De Wit, H. (2011). *Trends, Issues and challenges in internationalisation of higher education.* Amsterdam: Centre for Applied Research on Economics & Management, School of Economics Management of the Hogeschool van Amsterdam

Dussel, E. (2008). Philosophy of Liberation, the postmodern debate, and Latin American Studies. In M. Moraña, E. Dussel and C. A. Jáuregui (eds.) *Coloniality at Large: Latin America and the Postcolonial Debate* (pp. 335–478). Durham and London: Duke University Press

Enserink, M. (2018). Research on research. *Science*, 361:6408, 1178–1179

Fals-Borda, O. and Mora-Osejo, L. E. (2007). Beyond eurocentrism: Systematic knowledge in a tropical context. A manifesto. In B. S. Santos (ed.) *Cognitive Justice in a Global World: Prudent Knowledge for a Decent Life* (pp. 397–405). Lanham: Lexington Books.

Grosfoguel, R. (2007). The epistemic decolonial turn. *Cultural Studies*, 21:2-3, 211–223

Gu, M. D. (2020). What is'decoloniality'? A postcolonial critique. *Postcolonial Studies*, 23:4, 596–600

Guilherme, M. (2019a). Introduction: The critical and decolonial quest for intercultural epistemologies and discourses. In Intercultural Multilateralities: Pluridialogic imaginations, globo-ethical positions and epistemological ecologies (Special Issue). *Journal of Multicultural Discourse*, 14:1, 1–13

Guilherme, M. (2019b). Glocal languages beyond postcolonialism: The metaphorical North and the South in the geographical north and south. In M. Guilherme & L. M. T. M. Souza (eds.) *Glocal Languages and Critical Intercultural Awareness: The South Answers Back* (pp. 42–64). London and New York: Routledge

Guilherme, M. (2020a). Intercultural responsibility: Critical inter-epistemic dialogue and equity for sustainable development. In Leal Filho W., Azul, A.M., Brandli, L., Lange Salvia, A., Wall, T. (eds.) *Partnership for the Goals: Encyclopedia of the UN Sustainable Development Goals*, vol. 17. Cham, Switzerland: Springer Nature (https://link.springer.com/referenceworkentry/10.1007/978-3-319-71067-9_75-1)

Guilherme, M. (2020b). Intercultural responsibility: Transnational research and glocal critical citizenship. In J. Jackson, *The Routledge Handbook of Language and Intercultural Communication* (2nd ed., ch. 21). Abingdon, UK: Routledge

Guilherme, M. (forthcoming). Intercultural responsibility in conditions of conflict and crises: Glocademics in action. In P. Holmes and J. Corbett (eds.) *Intercultural Pedagogies for Higher Education in Conditions of Conflict and Crises: Culture, Identity, Language*. London and New York: Routledge

Guilherme, M. and Dietz, G. (2015). Difference in Diversity: Multiple perspectives on multi-, inter-, and trans-cultural conceptual complexities. *Journal of Multicultural Discourses*, 10:1, 1–21

Levinas, E. (1998). *On Thinking-of-the-Other: Entre Nous*. New York: Columbia University Press

Makoni, S. (2019). Conflicting reactions to chi'ixnakax utxiwa: A reflection on the practices and discourses of decolonization. *Language, Culture and Society*, 1:1, 147–151

Mignolo, W. D. (2000). *Local Histories/Global Designs: Coloniality, Subaltern Knowledges, and Border Thinking*. Princeton, New Jersey: Princeton University Press

Mignolo, W. D. (2007). Delinking. *Cultural Studies*, 21:2, 449–514

Mignolo, W. D. (2008). The geopolitics of knowledge and the colonial difference. In M. Moraña, E. Dussel and C. A. Jáuregui (eds.) *Coloniality at Large: Latin America and the Postcolonial Debate* (pp. 225–258). Durham and London: Duke University Press

Mignolo, W. D. (2009). Epistemic disobedience, independent thought and decolonial freedom. *Theory, Culture & Society*, 26:7–8, 1–23

Mignolo, W. D. (2018). The decolonial option. In W. D. Mignolo and C. Walsh (eds.) *On Decoloniality: Concepts, Analytics, Praxis* (pp. 105–226). Durham and London: Duke University Press

Mignolo, W. and Walsh, C. (2018). *On Decoloniality: Concepts, Analytics, Praxis*. Durham: Duke University Press

Miuht, G., Altbach, P. G. and De Wit, H. (eds.) (2017). *Understanding Higher Education Internationalization: Insights from Key Global Publications*. Rotterdam: Sense

Ndlovu, M. (2020). Well-intentioned but vulnerable to abuse. *Postcolonial Studies*, 23:4, 579–583

Robertson, R. (1995). Glocalization: Time-space and homogeneity-heterogeneity. In M. Featherstone, S. Lash and R. Robertson (eds.) *Global Modernities* (pp. 25–44). London: SAGE

Santos, B. S. (1999). Towards a multicultural conception of human rights. In S. Lash and M. Featherstone (eds.) *Spaces of Culture: City, Nation, World*. London: SAGE

Santos, B. de S. (2006). *Conocer desde el Sur: Para una cultura política emancipatória*. Lima: Universidad Mayor de San Marcos

Santos, B. de S. (2007). Beyond abyssal thinking: From global lines to Ecologies of Knowledges. *Review* 30:1, 45–89

Santos, B. de S. (2014). *Epistemologies of the South*. Boulder: Paradigm

Santos, B. de S. (2018). *The End of the Cognitive Empire: The Coming of Age of Epistemologies of the South*. Durham: Duke University Press

Spivak, G. C. (1988). Can the Subaltern Speak? In C. Nelson and L. Grossberg (eds.) *Marxism and the Interpretation of Culture* (pp. 271–313). Macmillan Education: Basingstoke

Spivak, G. C. (1999). *A Critique of Postcolonial Reason.* Cambridge, MA: Harvard University Press

Spivak, G. C. (2004) 'Righting Wrongs'. *The South Atlantic Quarterly*, 103 (: 2/3):, 523–581.

The South Commission (1990). *The Challenge of the South: The Report of the South Commission.* Oxford: Oxford University Press

Tiana-Ferrer, A. (2014). The impact of the Bologna process in Ibero-America: Prospects and challenges. In A. Teodoro and M. Guilherme (eds.) *European and Latin American Higher Education between Mirrors: Conceptual Frameworks and Policies of Equity and Social Cohesion.* Rotterdam: Sense Publishers

Veugelers, R. (2010). Towards a multipolar science world: Trends and impact. *Scientometrics*, 82, 439–456

Veugelers, R. & Van Bouwel, L. (2015). The effects of international mobility on European researchers: Comparing intra-EU and U.S. mobility. *Research in Higher Education*, 56, 360–377

Visvanathan, S. (2007). Between cosmology and system: The heuristics of a dissenting imagination. In B. Sousa Santos (ed.) *Another Knowledge Is Possible* (pp. 182–218). London: Verso

Walsh, C. (2012). 'Other' knowledges, 'other' critiques: Reflections on the politics and practices of philosophy and decoloniality in the 'Other' America. *Transmodernity: Journal of Peripheral Cultural Production of the Luso-Hispanic World*, 1:13, 11–27

Walsh, C. (2018). On decolonial dangers, decolonial cracks, and decolonial pedagogies rising. In W. D. Mignolo and C. Walsh (eds.) *On Decoloniality: Concepts, Analytics, Praxis* (pp. 81–98). Durham and London: Duke University Press

1 European governance and doctoral education

What is "higher" in higher education?

António M. Magalhães & Amélia Veiga

The chapter analyzes the political grammar that is shaping European higher education, particularly doctoral education, and focuses on the changes associated with the idea of producing "knowledge workers" in the so-called knowledge society and economy. European higher education is being reconfigured in the framework of the creation and development of the European Higher Education Area (EHEA). As the political grammar underlying these transformations is impinging on the reconceptualization of the university as a public institution and reconfiguring both the universities' internal life and their own mission, it is crucial to analyze its implications for doctoral education at the European level.

By looking at European policy drivers and governance instruments aiming to configure doctoral education, the argument in this chapter is that European higher education is being reshaped on the basis of a model of knowledge production to respond to the market needs, particularly a new labor force challenging higher education systems and institutions. This model assumes monocultural and mono-epistemological features by underlining the relevance of knowledge applicability and the alleged needs of the labor market. The major policy driver promoting the interaction between education and innovation is discursively associated in European Union (EU) policy to the ideograph of a "knowledge society" in tension with education as "an intentional set of processes aimed at producing worthwhile forms of human development, higher education has to be in the business of producing the most advanced forms of human development" (Barnett, 1997: 162).

The Bologna Process and the policies of international organizations such as the World Bank and Organization for Economic Co-operation and Development (OECD) were key in this reconfiguration. The training of knowledge workers became a key political and pedagogical driver in doctoral education and marked a political agenda focused on competences, skills, and measurement of learning outcomes assumed as an appropriate form of responding to the needs of the knowledge society and economy.

DOI: 10.4324/9781003225812-2

These drivers have also been reconfiguring higher education institutions themselves, questioning the Humbolditian, Newmanian, Napoleonic, and Deweyian models of the university under the rise of the "entrepreneurial" university (Clark, 1998). The reconceptualization of the university as a public institution is stimulating the shift from a republic of scholars to a stakeholders organization (Bleiklie & Kogan, 2007), which is taking place in a force field that is pressuring higher education dynamics.

The political grammar framing European higher education and the concern of making European higher education compatible, comparable, and economically relevant to the social and economic development of Europe convenes instruments that promote homogeneity rather than concern for differences and relations between forms of knowledge. The focus on the training of the knowledge worker – as a set of monocultural competences and skills for the envisaged knowledge society and economy – might end up reducing him/her to a set of competences and skills whose main characteristics are to be transferable and flexible. These drivers risk homogenizing cultural and socio-economic contexts and institutional and individual differences in the figure of the knowledge worker. A plurilingual, intercultural, and interdisciplinary perspective on transnational research risks vanishing by the incorporation of new epistemological possibilities in a movement that is expanding and globalizing.

On the basis of the analysis of documents issued by the European Commission and the communiqués by the Education Ministers involved in the Bologna process, and by other relevant stakeholders in European higher education policy, the chapter will critically approach the emphasis on making doctoral education a compilation of attributes, skills, and experiences to equip individuals for the labor market. Next, the chapter will look at selected governance instruments enacted at the European level to identify the features of the political grammar shaping doctoral education and influencing education worldwide. Interestingly enough, the development of the Ibero-American Knowledge Space can be seen mirroring the establishment of the EHEA (Tiana-Ferrer, 2014) underlining the dialectics between policies transfer and borrowing. The conclusion underlines the influence of these policy assumptions and drivers on the profile of doctoral education and the relevance of the knowledge worker, let alone what is "higher" in higher education itself. In other words, the linkage between research and education is key to keep a critical stance at the core of higher education (Barnett, 1997).

The Europe of Knowledge and Innovation

At the roots of the EHEA, there is a major policy driver promoting the interaction between education, research, and innovation, which is discursively associated in the EU policy rhetoric to the ideograph of a knowledge society and economy. In 1997, the notion of a Europe of Knowledge was introduced by the European Commission in putting forward the Agenda for 2000 to make "knowledge-based policies" (innovation, research, education, training) one of the EU pillars to raise the level of knowledge and skills of all Europe's citizens in order to promote employment (Commission of the European Communities, 1997). In 2003, knowledge policies were directed to the need for developing effective and closer cooperation between universities and industry, "gearing it more effectively towards innovation, new business start-ups and, more generally, the transfer and dissemination of knowledge" (European Commission, 2003). The European Commission made clear its commitment to promote (higher) education, research, and innovation in the creation of a "Europe of Knowledge" targeted by the Lisbon agenda (Commission of the European Communities, 1997).

Doctoral education became a European priority at the Ministerial Meeting of the Bologna Process in Berlin (Berlin Communiqué, 2003). In this communiqué, doctoral studies articulated the EU's two strategies for a EHEA and a European Research Area (ERA). These strategies, while meeting the priorities of the Lisbon agenda, feed the narrative of the European knowledge society. Doctoral degrees were to be more closely linked with careers in research and development and joint doctorates were to be developed more easily once obstacles to countries' recognizing each other's qualifications were expected to be removed.

In 2005, ministers stated that "As higher education is situated at the crossroads of research, education and innovation, it is also the key to Europe's competitiveness" (Bergen Communiqué, 2005). Since 2005, the Bologna Process was expected to enhance the relationship between higher education and research which, in turn, underpinned "higher education for the economic and cultural development of our societies and for social cohesion" (Bergen Communiqué, 2005). The Lisbon agenda assumes that "Modernisation is needed in order to face the challenges of globalisation and to develop the skills and capacity of the European workforce to be innovative" (European Commission, 2007: 1) while pointing out three areas of "possible reform" in higher education: curricular, governance, and funding. This was expected to have major consequences on the variety of doctoral programs and on the enhancement of provision of the third cycle to promote "the status, career prospects and funding for early stage researchers" as "essential preconditions for meeting Europe's objectives of strengthening research capacity and improving the quality and competitiveness of European higher education" (London Communiqué, 2009).

The European Commission's mandate for further action related to research was reinforced (Keeling, 2006) and it doubled the funds for research (7th Framework Programme), reaffirming the leading role of the European Commission in making Europe the most competitive economy and feeding discourses about the shift from basic to applied research. This endeavor has been promoting transnational research projects and the creation of the ERA where *gloacademics* play a key role. Furthermore, the focus on knowledge transfer contributes to blurring the distinction between research and its applicability. The assumption is that research brings about "add[ed] value to markets, governments and society" (European Commission, 2010). The emphasis shifts from research *per se* to research + innovation as mediated by knowledge transfer and, consequently, under the framework of the Bologna Process "Study programmes must reflect changing research priorities and emerging disciplines, and research should underpin teaching and learning" (Bucharest Communiqué, 2012).

Mainly after the Bergen meeting (2005), innovation became tightly articulated with education and research and simultaneously strengthened the link between research and innovation. While in the Berlin Communiqué (2003), the aim of preserving Europe's cultural richness and linguistic diversity was related to fostering of "its potential of innovation and social and economic development through enhanced co-operation among European Higher Education Institutions" (Berlin Communiqué, 2003: 2), the Bergen Communiqué (2005) clearly articulated innovation and education by assuming that "time is needed to optimise the impact of structural change on curricula and thus to ensure the introduction of the innovative teaching and learning processes that Europe needs" (Bergen Communiqué, 2005: 1). The articulation between research and innovation was key to molding the discourses that assumed innovation as the main political driver for economic growth (European Commission, 2010). Innovation is high on the EU agenda as reflected in the 2020 strategy:

> EU public policies should focus on creating an environment that promotes innovation (...). By improving conditions and access to finance for research and innovation in Europe, we can ensure that innovative ideas can be turned into products and services that create growth and jobs
>
> (Bucharest Communiqué, 2012).

and EU ministers of education recognized the need to improve "cooperation between employers, students and higher education institutions, especially in the development of study programmes that help increase the innovation, entrepreneurial and research potential of graduates" (Bucharest Communiqué, 2012).

The EU metaphor of the knowledge triangle brings forward the relationships between education, research, and innovation underlining the dominance of the latter in its articulation with the other two vertexes of the triangle. This metaphor emphasizes the role of innovation in configuring the relationship between the university's "first" and "second" missions, i.e., education and research.

Innovation refers to the impact of higher education systems on economic development, on the enhancement of competitive advantages of regional systems and on the generation of skills for that purpose. Knowledge for economic and social development involves the strengthening of its impact on the basis of international knowledge production and cooperation, and the enhancement of regional systems with a focus on their competitive advantage. The idea of innovation and its articulation with research and education are configuring the landscape of doctoral education. The European governance system, while consolidating the role of the European Commission as a governance supranational body, contributes to legitimizing national discourses and decisions on higher education issues.

The articulation between research and innovation is based on the presupposition that "Europe has world-class researchers, entrepreneurs and companies" and that "Europe's research and innovation performance needs to be boosted to master the many challenges ahead and keep its place in a fast changing world" (European Commission, 2010). From the perspective of the European Commission, the four things that enable innovation are: human resources, open and excellent research systems, finance, and support. Human resources development is linked to the

> importance of research and research training and the promotion of inter-disciplinarity in maintaining and improving the quality of higher education and in enhancing the competitiveness of European higher education more generally
>
> (Berlin Communiqué, 2003).

This is expected to have major consequences on the variety of doctoral programs and on the enhancement of provision of the third cycle to promote "the status, career prospects and funding for early stage researchers" as "essential preconditions for meeting Europe's objectives of strengthening research capacity and improving the quality and competitiveness of European higher education" (London Communiqué, 2007: 4–5).

Innovation aims to create in Europe world-class performers in science, meaning attractive careers for researchers, high-standard training, open access to research results, and cross-border mobility. Indicators of measurement of innovation performance are related to doctoral education and the number of new doctorate graduates, to international scientific co-publications, and to R&D expenditure in the public and business sector. These indicators illustrate how this emphasis on the expected economic effects (e.g., employment, services, and products in knowledge-intensive activities) potentially influences higher education.

Innovation assumes centrality in the knowledge triangle impinging on the nature of the teaching and learning processes and fixes the meaning of knowledge dissemination, pointed out as quintessential to the pursuit of an "innovative" workforce. In other words, knowledge dissemination is expected to provide new graduates "with the proper skills regarding the management, protection and exploitation of knowledge and intellectual property" (European Commission, 2005: 15). Actually, the ministers in London urged "institutions to further develop partnerships and cooperation with employers in the ongoing process of curriculum innovation based on learning outcomes" (London Communiqué, 2007: 6). This translates into the significance of issues "such as transparent access arrangements, supervision and assessment procedures, the development of transferable skills and ways of enhancing employability" (London Communiqué, 2007: 5).

The European University Association also considered that doctoral programs are "a crucial source of a new generation of researchers and serve as the main bridge between the European Higher Education and Research Areas" (Reichert & Tauch, 2005: 7). A report by this Association, while assuming that doctoral programs have become an important part of EU strategies and the Bologna Process, recognizes that "the reforms of doctoral education are proceeding at varied paces" (Reichert & Tauch 2005). The Lisbon agenda objectives, the EC research policies, and the Bologna Process had impacts on the development of the doctorate (called the third cycle after the bachelor's and master's "cycles"), either by inducing clear adaptive changes to the structure of the doctorate (e.g., Portugal) or by accelerating the reform processes (e.g., France, Germany). Actually, "The organisation of doctoral programmes displays a large diversity not only across different countries in Europe, but also across universities within the same country and across faculties within the same university" (Reichert & Tauch, 2005: 12). Not only do countries have diverse legal frameworks and regulations, but, where universities enjoy a greater deal of autonomy, doctoral programs are the university's own responsibility. This diversity in the doctorates, however, is used in many cases, at least at the discursive level, as a doorway for the development of European convergence efforts. Common and clear guidelines and regulations with regard to access, supervision, and evaluation are expected to enhance the transparency and comparability, which are assumed to be the basis of EU citizenship itself.

Based on the assumption that education in higher education is to be tightly related with the economic fabric, employability is conceived of as the potential to participate in the knowledge economy and thus envisages the adequate competences and skills (e.g., Leuven communiqué). Concomitantly, mobility is an element of knowledge dissemination, particularly in doctoral education, as the promotion of mobility is expected to equip the graduates with innovation skills to cope with the risk features of knowledge societies and their labor markets. As underlined by the Yerevan communiqué, "greater mobility of students and staff fosters mutual understanding, while rapid development of knowledge and technology, which impacts on societies and economies, [and] plays an increasingly important role in the transformation of higher education and research" (Yerevan Communiqué, 2015: 1). These assumptions reflect on curricular reforms and, ultimately, on the establishment of standards, guidelines, and procedures in education. The concern with the effectiveness in achieving learning outcomes was framed by the concern of economic relevance of innovations, risks diluting the critical dimensions of doctoral education, and failing the opportunity to engage in the education of the global citizen.

In turn, the emphasis on learning outcomes made of skills and competences for innovation while responding to the mandate for education systems to develop the right mix of skills (European Commission, 2010 and OECD Innovation Policy Platform) shows the extent to which the European Commission and the OECD share this political grammar. Actually, the report by the *Expert Group on New Skills for New Jobs* emphasized that education and training "must be underpinned by transversal competences, especially digital and entrepreneurial competences, in order to both encourage initiative rather than simple reproduction of received knowledge and to better adapt to learners' and employers' needs" (European Commission, 2010: 7), and the *Innovation Policy Platform (IPP)*, developed by the OECD and the World Bank, underlined the need "to rebalance the emphasis between content knowledge and other skills such as creativity, communication, teamwork (...)". According to these organizations, the acquisition of innovation skills is based on: (i) disciplines that are expected to equip students with skills that matter for innovation: technical skills, skills in thinking and creativity, and behavioural and social skills; (ii) pedagogies that must be active based on problem-based learning, cooperative learning, metacognitive learning, sometimes enhanced by information and communication technology and on interdisciplinary approaches focusing on design thinking to foster skills for innovation and (iii) new assessment instruments focusing on competences rather than mere knowledge, and (iv) international mobility of students, faculty, programs, and institutions, introduced as a means to foster skills for innovation in the globalized economy. Doctoral education is deemed to develop skills related to professional and career management and networking, self-management (preparation and prioritization, responsiveness to change, work-life balance), self-reflection, responsibility and integrity, working with others (collegiality, team-working, supervision, mentoring, equality and diversity), communication, and dissemination (communication media and publication). The call for innovation skills is promoting the knowledge worker and aims to equip individuals for the labor market in a compilation of attributes, skills, and experiences potentially limiting the scope of higher education to its economic relevance, or in Barnett's terms, sacrificing the formation of the *critical-self* to the *corporate-self* (Barnett, 1997).

The European discourse has been influencing university dynamics and the concepts and designs associated with doctoral education. Guidance has been issued on methods of recruiting researchers and employment and working conditions. The approach to doctoral training, research, and careers is now to include "wider employment-related skills, the structuring of training, the quality of supervision and the funding of doctoral programmes and candidates" (Jamieson & Rajani, 2006: 3).

The Political Goals Toward Convergence: Between the Three-Cycle Structure and Doctoral Schools

The governance instruments developed at the European level (e.g., Open Method of Coordination) reflect a pragmatic approach by European institutions toward the way higher education should be governed. This is of importance as, as argued by Lascoumes and Galès (2007), policy instruments impinge on the nature of the policy itself. The use of "soft" law used in the implementation of Bologna's action lines, i.e., the credit system, the Diploma Supplement, and the degree structure, have been framing the reforms in the signatory countries. Reforms of doctoral education convened the convergence of degree systems and degree structures, and the setting up of doctoral schools.

The discourse of the Europe of Knowledge challenged national and institutional contexts in framing the reforms of doctoral education. For instance, in France, since the contractual policy initiated in 1983, the doctoral schools are acknowledged as the locus of "vocational experience in research" (Musselin & Paradeise, 2009: 43). In Norway, the "new" doctoral degrees (i.e., the structured doctoral programs) organized on the basis of doctoral programs, and the structuring of doctoral education in general, can be considered as a state's objective to increase the formalization of the system aimed at making doctoral programs more efficient and predictable (Bleiklie, 2009). In the United Kingdom, even though doctoral education reforms developed without being directly concerned with the European processes (Kehm, 2009), since the 1990s a utilitarian view of research policy was visible by promoting close connections between science/universities and industry. The political promotion of "big science" by means of inter-institutional cooperation to create critical mass across clusters of universities within particular subjects was also crucial.

The extent to which European governance instruments induce the centrality of the training of knowledge workers, limiting their profile to a set of competencies and skills, is under scrutiny. Additionally, and at a later stage, it is important to raise the question of whether the drivers reforming doctoral education risk homogenizing cultural and socio-economic contexts, turning a blind eye to institutional and individual differences in the figure of the knowledge worker.

By 2010, the development of the Bologna Process included the introduction of a three-cycle structure. The Bologna degree structure was introduced mainly to respond to national legal frameworks. However, it did not always lead to meaningful curricular renewal (Sursock, 2015: 69) and there are variations in understandings of what constitutes a bachelor's, a master's, or a doctoral degree (ESU, 2015). The most typical duration of full-time doctoral programs is three years; however, the duration of doctoral studies reaching beyond this time is also relevant (European Commission/EACEA/Eurydice, 2015). Additionally, while

> the proportion of structured doctoral programmes is growing, the traditional supervised doctoral studies are still the most widespread. In 16 countries all doctoral training follows such a traditional model and in another nine countries over 70% of programmes follow the traditional approach. Professional doctoral programmes are not yet widespread. Only Belgium (Flemish Community), Denmark, Ireland and the United Kingdom have 2-5% professional doctoral programmes
>
> (European Commission/EACEA/Eurydice, 2015: 65).

Furthermore, there are countries that have developed doctoral qualifications on the basis of the qualifications frameworks (e.g., France, Italy, Norway, and the United Kingdom (Scotland)) (Implementation Report) and the use of ECTS in doctoral studies spread widely across higher education systems (Implementation Report). Diversity in doctoral studies prevails while the theme went high in the EU higher education agenda envisaging the convergence of degree systems and structures, bringing to the fore the tension between the European dimension toward convergence and the preservation of diversity in doctoral studies.

At the national level, the Netherlands and Norway, largely using the Bologna momentum, moved closer to the European concern with doctoral education/training aiming to provide transferable skills, adding to the traditional one-to-one apprenticeship multiple supervision and consolidating doctoral schools and structured programs. At different paces and rates, for instance, Germany, Italy, Portugal, and Switzerland have been enacting policies aimed at transforming doctoral education in line with EC injunctions. On the one hand, the academic cultures and institutional ethos, with important nuances among the countries, are making it difficult to attain transparency and comparability; on the other hand, time is needed to know how these changes will evolve. However, the reorganization of doctoral education/training in structured programs is emerging as a commonality against the traditional doctoral studies.

The existing models of doctoral education in the EHEA (Kehm, 2009) reflect practices of academia and encompass the research doctorate, the taught doctorate, PhD by published work, the professional doctorate, the practice based doctorate, the "new route" doctorate, and joint doctorates. At the same time, there was an effort toward convergence visible in the recognition of "ten basic principles" for the future development of doctoral programs (European University Association, 2005) that informed the formulation of recommendations for the Bergen Ministerial meeting (2005). These principles are reshaping doctoral training, on the one hand, underlining the need to respond to the demands of an employment market wider than academia and the need to achieve critical mass drawing on different types of innovative practice being introduced in universities across Europe; on the other hand, emphasizing the need to meet the challenge of interdisciplinary training and the development of transferable skills. In line with this, some countries have established guidelines, codes, and regulations which include the rules of recruitment, supervision, exams, evaluation, and defense of the final thesis driven by national research policies (e.g., Switzerland has adopted the European Charter of Researchers and the Code of Conduct for the Recruitment of Researchers; and France the *Pacte pour la Recherche*).

In spite of the transparency efforts toward convergence, diversity in doctoral studies remains. The heterogeneity of doctoral models does not derive, at least directly, from the Bologna process since their differentiation started in the early 1980s. The adoption by the EU of the EQF (2008) and its translation into the National Qualifications Frameworks have been providing a basis for comparability and transparency. However, the vagueness of the formulation about the nature of the competences and learning outcomes does not seem enough to bring more comparability between the diverse doctoral models. According to B. Kehm, with the increasing national and international competition for doctoral students and along with the tendency toward the vertical stratification of European universities, "the tension between diversity and transparency tends to be solved by substituting the horizontal dimension of differentiation into a hierarchical order" (Kehm, 2009: 233). The three-cycle system and doctoral education have been evolving within the tension between the EC call for convergence and national dynamics and institutional diversity. However, the political grammar framing European higher education is directing the development of doctoral education toward the training of the knowledge worker; i.e., to promote a set of universal competences and skills for the knowledge society and economy. On the one hand, the figure of this knowledge worker appears to be conceived beyond national and cultural differences as if it is a taken-for-granted figure; on the other hand, the training models appear to be reconfigured under the drive of doctoral schools. These developments reflect the loss of national hegemony as universities worldwide are tending to adopt a monocultural and mono-epistemological model of knowledge production that aims to respond directly to the market.

Despite the drive toward convergence, doctoral schools have been introduced as a means to deal with diversity as far as they are directed toward interdisciplinary training and the development of transferable skills. The enactment of this political grammar has resulted in the "rapid expansion of doctoral schools and improved supervision and training of doctoral candidates" (Sursock, 2015: 69). Actually, in 2014,

> 32 systems state that they have doctoral schools – an increase from 30 in 2012. There are 16 systems which do not have doctoral schools (...). In 12 of the countries where doctoral schools exist, 1-25% of doctoral students study within such structures. There are 19 systems where the majority of students study in doctoral schools, including seven where all students study in such a framework.
>
> (European Commission/EACEA/Eurydice, 2015: 66).

Governance reforms developed across Europe under the influence of New Public Management at different rates and paces must be taken into account when looking at the development of doctoral schools. Actually, at the institutional level, the dynamics involving the setting up of doctoral schools goes hand-in-hand with governance, finance, and curricular reforms. For instance, in Portugal the legal framework established in 2007 has strengthened the managerial bodies to the detriment of collegial bodies, the centralization of decision-making processes, and the presence of external stakeholders at central and faculty/school/department levels. Additionally, five public HEIs opted for the foundational model enabling them to be ruled by private law. Furthermore, the establishment of the A3ES (Agency for Assessment and Accreditation of Higher Education) introduced the accreditation criteria that makes compulsory the existence of doctoral programs associated with research centers classified as Very Good or Excellent, which in turn played a role in supporting the decision to create doctoral schools. This reflects the dissemination and hegemony of a scientific model based on the evaluation culture stemming from entrepreneurial forms of organization of academic work.

These schools as the privileged model for doctoral training appear as a convergence trend echoing the EU political grammar and its goals. The training of knowledge workers with skills and competences for the labor market is shaping institutional and academic dynamics and influencing the concept of doctoral education. Diversity is dealt with under organizational perspectives rather than on effective differentiation of education models at the doctoral level. One might question the extent to which doctoral schools contribute to the blurring of the differences between doctoral education in the creation of the figure of the knowledge worker hindering the broader goal of individual's formation wider than the economic concern. This might indeed ignore the need for recognizing institutional, academic, disciplinary, and individual differences as a way to deal with diversity and convergence. Furthermore, the setting up of doctoral schools appears to convene both the European drivers toward economic relevance of education while assuming implicitly epistemological supremacy driven by values centered on competition and value for money.

Conclusion

The relationship between knowledge and doctoral education has been changing. Education and the "higher" of higher education in the context of massification are framed by economic competition and the need to meet changing labor market requirements and expectations. The centrality of knowledge in the modern ideal of higher education is shifting to competences and learning outcomes for the knowledge society and economy. The influences of policy assumptions and drivers on the profile of doctoral education are inducing the relevance of the knowledge worker, let alone what is "higher" in higher education itself. As argued, these drivers and assumptions in their taken-for-grantedness tend to naturalize the hegemonic epistemology both within and without Europe and legitimize the homogenization of research prospects and related education perspectives.

In the European context, these policy assumptions and drivers are focusing on output-based approaches to education underlying the instrumentality of knowledge and linking it to labor market and economic needs. The discourses and implementation of doctoral schools are inducing the replacement of the formative role of knowledge, particularly in the world of work. In the labor market, competences and qualifications, set out as learning outcomes to be measured, are assumed to be the basis of "employability" and a common grammar for the use of higher education external stakeholders, namely for employers.

The reconfiguration of educational concepts and structures aims to engage higher education institutions and their constituencies with the ultimate goal of promoting mobility and transparency between higher education systems and institutions. Additionally, the potential these transformations have with regard to the flexibility of training trajectories allegedly responds to massification and the need to recognize diversity. An overemphasis on the applicability of knowledge can be read as a trend to neglect the formative role that knowledge has *per se*. This might bind the educational mission and strategies of higher education to strict economic and vocational roles. Hence, the focus on convergence, namely by means of setting up doctoral schools, may not cope with the multitude of learning needs, paths, and projects involving the diversity of doctoral students.

Last but not least, and beyond the diversity of doctoral students, institutional diversity is at stake. As questioned by Brennan: "how far will the notion of an education that is 'higher' than other forms continue to be viable within differentiated knowledge organisations of the knowledge society?" (Brennan, 2012: 201). This remains to be seen.

References

Barnett, R. (1997). *Higher Education: A Critical Business*. London: The Society for Research into Higher Education.

Bergen Communiqué. (2005). *The European Higher Education Area - Achieving the Goals*. Bergen. Retireved at http://ehea.info/page-ministerial-declarations-and-communiques.

Berlin Communiqué. (2003). *Realizing the European Higher Education Area*. Berlin. Retrieved at http://ehea.info/page-ministerial-declarations-and-communiques.

Bleiklie, I. (2009). Norway: From Tortoise to Eager Beaver. In C. Paradeise, E. Reale, I. Bleiklie, & E. Ferlie (Eds.), *University Governance: Western European Comparative Perspectives* (pp. 127–152). Dordrecht: Springer.

Bleiklie, I., & Kogan, M. (2007). Organization and Governance of Universities. *Higher Education Policy*, *20*(4), 477–494.

Brennan, J. (2012). Is There a Future for Higher Education Institutions in the Knowledge Society? *European Review*, *20*(02), 195–202. doi:10.1017/S1062798711000421.

Bucharest Communiqué. (2012). *Making the Most of Our Potential: Consolidating the European Higher Education Area*. Bucharest. Retrieved at http://ehea.info/page-ministerial-declarations-and-communiques.

Clark, B. (1998). *Creating Entrepreneurial Universities, Organizational Pathways of Transformation*. Pergamon Press. Retrieved at http://ehea.info/page-ministerial-declarations-and-communiques.

Commission of the European Communities. (1997). *Towards the Europe of Knowledge*. Brussels.

European Commission. (2003). *The Role of Universities in the Europe of Knowledge*. Brussels: Pergamon Press.

European Commission. (2005). *Modernising Education and Training Systems: A Vital Contribution to Prosperity and Social Cohesion in Europe*. Brussels: European Commission.

European Commission. (2007). *From Bergen to London - The Contribution of the European Commission to the Bologna Process*. Brussels: European Commission.

European Commissio n. (2010). *Communication from the Commission - Europe 2020 - A Strategy for Smart, Sustainable and Inclusive Growth*. Brussels: European Commission.

European Commission/EACEA/Eurydice. (2015). *The European Higher Education Area in 2015: Bologna Process Implementation Report*. Luxembourg: European Commission/EACEA/Eurydice.

European University Association. (2005). *Conclusions and Recommendations*. Bologna Seminar on "Doctoral Programmes for the European Knowledge Society". Salzburg.

European Students' Union (ESU). (2015). *Bologna with Student Eyes - Time to Meet the Expectations from 1999*. Brussels:European Students' Union.

Jamieson, I., & Rajani, N. (2006). University Positioning and Changing Patterns of Doctoral Study: the case of the University of Bath. *European Journal of Higher Education*, *42*(3), 363–373.

Keeling, R. (2006). The Bologna Process and Lisbon Research Agenda: The European Commission's Expanding Role in Higher Education Discourse. *European Journal of Education*, *41*(2), 203–223.

Kehm, B. (2009). New Forms of Doctoral Education and Training in the European Higher Education Area. In B. Kehm, J. Huisman, & B. Stensaker (Eds.), *The European Higher Education Area: Perspectives on a Moving Target* (pp. 223–241). Rotterdam: Sense Publishers.

Lascoumes, P., & Galès, P. L. (2007). Introduction: Understanding Public Policy through Its Instruments - From the Nature of Instruments to Sociologiy of Public Policy Instrumentation. *Governance: An International Journal of Policy, Administration and Institutions*, *20*(1), 1–21.

London Communiqué. (2007). *Towards the European Higher Education Area: Responding to Challenges in a Globalised World*. London. Retrieved at http://ehea.info/page-ministerial-declarations-and-communiques.

Musselin, C., & Paradeise, C. (2009). France: From Incremental Transitions to Institutional Change. In C. Paradeise, E. Reale, I. Bleiklie, & E. Ferlie (Eds.), *University Governance: Western European Comparative Perspective* (pp. 21–49). Dordrecht: Springer.

Reichert, S., & Tauch, C. (2005). *Trends IV: European Universities Implementing Bologna*. Brussels: European University Association.

Sursock, A. (2015). *Trends 2015: Learning and Teaching in European Universities*. Brussels: European University Association.

Tiana-Ferrer, A. (2014). The Impact of the Bologna Process in Ibero-America: Prospects and Challenges. In António Teodoro and Manuela Guilherme (Eds.), *European and Latin American Higher Education Between Mirrors* (pp. 125–138). Rotterdam: Sense Publishers.

Yerevan Communiqué. (2015). *Yerevan Communiqué*. Yerevan. Retrieved at http://ehea.info/page-ministerial-declarations-and-communiques.

2 European and Latin American researchers between mirrors

The internationalization of higher education for social cohesion and equity

Manuela Guilherme & António Teodoro

Introduction

This chapter provides an example of the functioning of a large transnational and intercontinental network and project Ibero-American Research Network in Education Policies (RIAIPE3) by pausing upon a number of participants' views, through their own voices, about the collaboration between European and Latin American researchers. This project was previous to the *Glocademics* project and we believe that our experience in the coordination of this project, moreover based on the real voices of a great number of its experienced researchers, may complement the descriptions and reflections of the *Glocademics* project comprehended in this book. The focus here is though on higher education institutions (HEIs) networks and research on education. The beginning of the RIAIPE3 project preceded a 2013 "Communication from the Commission to the European Parliament, the Council, the European Economic and Social Committee and the Committee of the Regions" about "European higher education in the world" which starts by declaring that "Education, and in particular higher education, is at the heart of the Europe 2020 Strategy". The project we address in this chapter is the *Inter-University Framework Program for Equity and Social Cohesion Policies in Higher Education* (RIAIPE3), involving 22 Latin American HEIs and 8 European ones, attempted to move beyond the study of higher education in abstract terms, e.g. "global internationalisation of higher education". For that purpose, it aimed to critically clarify the different implications that the common use of general concepts may convey, such as democracy, equity, governance, social cohesion, etc., and, simultaneously, share different types of good practice that the participants had been experimenting as well as those they created along the three-year cooperation.

However, our main objective here is not to present a description of the project or the narration of its activities, which will only act as the backdrop of the critical analysis of the scientific development of the network and the impact of the project activities in the participant institutions and

DOI: 10.4324/9781003225812-3

their students' communities, based on the video-recorded testimonies of the project researchers themselves. To start with, we engage in a critical discussion of the broader topic of the so-called internationalization of higher education, followed by some reflections on network organization, before we finally convey the final reflections on this 3-year commitment voiced by its actual actors.

The Internationalization of Higher Education

Literature on Higher Education Internationalization (HEI) is rather recent but blossoming. The concept is neither new nor homogeneous (Hudzik 2015), although it had not been so broadly planned as it is today, it has always been mainly driven by a colonial spirit, either over or covert. Hudzkik calls it "comprehensive internationalization" as it involves more complexity than the term "internationalisation of higher education" may seem to have involved, since it entails "the blending of diverse cultures and epistemologies" which the author traces back to thousands of years ago when "intellectual hubs drawing mobile scholars" were already scattered around the globe, namely in Athens and Italy as well as in China, Persia, India, and Alexandria (2015: 7). According to the author, academic internationalization nullifies the global-local divide with regard to knowledge exchange, since each group dynamics is unique and timely, as it shall be evident to have happened in the RIAIPE3 network addressed below.

Currently, HEI studies have concentrated their attention mainly on policies, both at national and supranational levels, student mobility programs, curriculum change, etc (Kehm and Teichler 2007). Discussions have ranged from economic globalization, the world labor market to the knowledge society, and the information age. Comparative studies have been dichotomous between comparative analyses, on similarities, and contrastive analyses, building on differences. Linguistic and intercultural complexities, analyzed both diachronically and synchronically, have been absent and bibliographic sources have been almost exclusively in English, as most authors in the field are now starting to acknowledge. However, HEI studies have proliferated worldwide and created a bibliographic bulk that is now considered specific scientific expertise as related to "the emergence of both new trans-national spaces of policy and new intra-national spaces of policy" (Ball 2016: 549). Ball reminds us that the so-called "internationalisation of higher education" has overcome and moved beyond the national paradigm, having had the national unit as the sole measurement unit, and he grounds his reflection on "the use of *network ethnography*". According to him "networks are not just a set of connections between sites but are also a history of on-going effort, and are animated by social relations and performance" (p. 550).

In this aspect, literature has highlighted what Rumbley and Altbach called "the crucial nexus" between "internationalisation's global and local dimensions" (2016) while, in the same body of work, other authors also make statements about the increasing prominence of the impact of the local in the re-significance of internationalization at the global level. Although the works mentioned above have initiated a critique of the path, local internationalization has taken by mentioning the excessive role of the English language, with either pro or con arguments, and the need for intercultural competence, the discussion still remains at a still too incipient level (Altbach 2013). De Wit has been more precise when he singles out, among his "nine misconceptions concerning internationalisation", as number one, the idea that "internationalisation is education in the English language" while he also concludes that, although internationalization and interculturality in higher education have been kept apart, there is a growing awareness of the "need to further connect these two subjects" (2011: 11/19). The RIAIPE3 project/network examined below, was a good example of an active trilingual network (Portuguese, Spanish, and English) with other subsidiary active languages (French and Italian, not to mention Dutch and Catalan), while remaining attentive to Latin American indigenous languages. This project/network's linguistic wealth reflected its intercultural amplitude and idiosyncrasies. Not only was *interculturalidad*/interculturality a way of life and of working but it was also the specific focus of some events and publications (Teodoro et al. 2013).

The complex relationality between the global and the local has been critically examined by some authors, as it was mentioned in the Introduction to this book. Urry argues for a "global complexity" within which there is no linearity between the global and the local (2005). Under the title "La globalización imaginada" (Imagined globalization), Canclini explains that "only some politicians, investors and academics think about the whole world, in a kind of circular globalization, and they are not even the majority in their professional fields. The rest imagines tangential globalizations" (my translation) (2003: 10). Santos refers to "the turbulence of scales" which "manifests itself through a chaotic confusion of scale among phenomena" (2014: 82). In sum, there is not one globalization, instead several globalizations, as Santos (2014) argues in his theorization of the "Epistemologies of the South".

In addition, a group of scholars (e.g. Castro-Goméz and Grosfoguel, 2007), who are in one way or another linked to Latin America (LA), have been theorizing about a "de-colonial turn" (*el giro decolonial*) that Mignolo (2007) defines as an epistemic decolonization, a kind of thought that releases and opens a critical border thinking. It is not possible to deal with intercontinentalization or internationalization of higher education, or even its transnational and intra-national dimensions, without taking a decolonial approach to any of the related issues, be it institutional governance, student access or success, curriculum management or social responsibility, equity or social cohesion, without understanding the implications of "decoloniality", that is, "thinking otherwise"

(Mignolo and Walsh 2018). Mignolo also claims that, "we need to think seriously about the processes by which languaging and the allocation of meanings to groups of people ... are being relocated ..." (2000: 236).

This situation can only be reversed by what Santos (2007) proposed as "post-abyssal thinking" that naturally flows into an "Ecology of Knowledges". Nevertheless, "the ecology of knowledges does not entail accepting relativism" (Santos 2014: 190), hence, this is not an assumption only valid for the relationships between Portugal and Brazil or Spain and the Spanish-speaking Latin American countries, it is as intercontinental as it is all too evident intra-nationally everywhere. It shouts between North and South, North and South America inclusive, also between northern South America and southern South America, northern and southern Europe, and East-West. In fact, this is a process to be undertaken together, for in Visvanathan's words: "We fought the West but the West, like the Orient, was not just out there but something within ourselves" (2007: 182). Spivak is even sharper in her words: "Elite 'postcolonialism' seems to be as much a strategy of differentiating oneself from the racial underclass as it is to speak in its name" (1999: 358). All in all, it becomes clear that globality in decolonial times requires that every participant reappraises each one's position and acknowledges that "all ignorance is ignorant of a certain kind of knowledge, and all knowledge is the overcoming of a particular ignorance" (Santos 2014: 188). Finally, researchers from The Centre for the Study of Developing Societies (CSDS) in New Delhi claim that they "move from the position of a critic of Western theory to that of one which composes and assembles new theory from different sources and different histories" (Banerjee et al. 2016: 42). To conclude, all in all Santos argues that: "The question does not imply that Europe needs to take lessons but rather engage in a new conversation with the world, a process of reciprocal learning based on more horizontal relations and mutual respect for differences" (Santos 2017b: 174).

Among other authors, Santos has called for cognitive justice. Hence, education is primordial and here the focus is the university. In *Decolonising the University*, Santos (2017a) once again reminds his readers that the oldest universities and elite learning centeres were not only placed in Europe but also in Mali and Egypt, for example. In fact, the first universities in the Americas, were founded by Jesuits and followed European models, however, the higher education institutions in LA have developed according to their historical, political, and sociological contexts mainly through the last century and have improved strategies to respond to their social responsibilities, more than the European ones, despite also suffering from the marketization that is negatively pressing university knowledge creation around the world (Red Iberoamericana de Investigaciones en Políticas Educativas 2012). Santos (2017a) also calls our attention for emergent universities in LA, which are contributing for a "pluriversity of knowledges", as he names it, for example, popular universities in Brazil, linked to social movements, some of which were created during the Lula era, as well as "intercultural" or "indigenous" universities, for example in Mexico and Colombia, both of which deserved the attention of the RIAIPE3's project (Guilherme and Santamaria 2015; Guilherme and Lourenço 2015; Guilherme and Dietz 2017). Networking between different types of universities makes for an enriching process of exchange, the more different the more the learning potential once collaboration remains horizontal and learning becomes reciprocal. However, there is a risk that attention is diverted from local problems; therefore, knowledge creation and experience gathering should remain at the grassroots level and never forget the contextual challenges to be resolved (Leite and Pinho 2017).

The Relevance and Operationalization of Research Networks

The creation of international networks has contributed decisively to the process of construction and consolidation of the social sciences, as a whole, and particularly of education. These processes can be explained within the framework of the so-called society of knowledge and social reflexivity. Both phenomena – the configuration of a networked society and the growing access to information – are the product of an age when knowledge and social media producers live, concurrently, the actual experiences which enable the construction of these networks and this shared knowledge.

Manuel Castells devoted a substantial first volume of his trilogy on the contemporary world to the description of the Networked Society (Castells 1996). According to the Spanish sociologist, the concept of networked society marks the emerging social structure within the information age, gradually replacing the society of the industrial age. The networked society is global, but features specific characteristics in each country, depending on its history, its culture and its institutions. This is a networked structure as a predominant form of organizing all activity. The networked society does not emerge because of technology, but without the information and communication technologies (ICT) it could not exist. In the last 20 years, the concept has expanded, and it now describes almost all social practices, including sociability or socio-political mobilization, based on the internet and on mobile platforms.

While seeking to explain contemporary social phenomena, Anthony Giddens developed the concept of social reflexivity (Giddens 1994). According to the British sociologist, reflexivity concerns knowledge disseminated outside the scope of experts, outside what he calls "expert systems", precisely by means of the information and communication networks which were set up thanks to the development of computation technologies and the internet. In other words, thanks to these new communication media, information, even that which does not come from expertise, has become accessible to the internauts, going beyond the boundaries of the universes of the insiders (researchers and scientists).

We know that access to information alone does not automatically produce a social network by its users. Besides access, it is necessary for the network actors to appropriate the information.

The Illuminist vision of knowledge dissemination had a single direction: it went from its producers (researcher, scientist, thinker, intellectual) to the receiving mass. While, on the one hand, eighteenth-century Western thinkers developed rationalist optimism – the ability to know oneself and change everything for knowledge –, as well as encyclopedic optimism – it is enough to know everything to command and transform everything, on the other hand, they also developed a kind of gnoseological pessimism of the majority and, in this way, the epistemological elitism commonly known as "vanguardism".

In several of his works (e.g, *The Politics of Education*, 1985), Paulo Freire referred to the evil of "vanguardism", but he also warned about the dangers of its reverse, that of "basism". Not everything that is drafted by the intellectual elites solves the problems faced by humanity, but likewise not everything that comes from the social base is better, or is alternative knowledge to the hegemonic knowledge, also because, for the most part, the masses "host" what is dominant and read the world from its perspective.

The social knowledge networks restore the possibility of gnoseological democracy, already hinted by the concept of reflexivity, and then expanded into the concept of "dialogic democracy" (Giddens 1994). In a more reflexive and globalized social order, there is the need to enhance more radical forms of democratization. Dialogic democracy is part of a democratization process, and it consists in the creation of a public arena where controversial issues can be resolved by dialogue rather than by pre-established forms of power (Arendt 1958; Habermas 1962). Both in daily activities and in social organizations, or even in the formal political sphere, individuals shape social practices and act together to find alternatives and overcome their problems and shortcomings collectively and reflexively.

Other concepts may emerge when we speak of social networks: shared knowledge and experiences, participative democracy, interlocution, alliances, collective (or trans-individual) action, connecting links, communicational process, interweaving the culture of encounter, among many others that can be listed. The essence common to all of them is unity in diversity, in thought and in action, in theory and in practice (Freire 1985).

We may then define a social network as a set of connections, voluntary or involuntary, of people or groups, who do not have the same borders of action but who present themselves as a structure which, in certain contexts, acts in the pursuit of common goals (Teodoro, Torres and Romão 2014). The social network is a kind of response to social fragmentation, sometimes gaining predominance as an alternative, while others as mediation between the state and society, between the public sphere and the private sphere. In the whole social network, rules of complementarity and reciprocity are defined, rules that are not always explicit, but rather implied in the common interests of specific contexts. As it happens in general communication, besides the sender, the receiver, the common code, the media and the message, which is unknown – if the content of the message were known beforehand by the speakers, there would be no communication –, in the interaction of social networks the different competencies complementing one another are only apparent as well as their interests which are, to some extension, instigated by the action of the other. Hence, the bargaining, the agreements and the adjustments that enable entry and permanence in the networks, are of different types, requiring mutually complementary expertise in order to mediate the challenges put forward by actual circumstances, and eventually overcome them.

The networks emerged in the sociological literature in the 1980s, especially when the political coordination of western societies began to shift away from the corporative organizations and surrender to the pressure of the so-called job-market. Networks arose as a kind of response to that shift, since the "market" has, to start with, being unable to respond to social needs and justice. Above all, social networks tend to compensate a deficit in representation and political coordination, previously provided by the state and to which the market is unable to respond fully or efficiently. To a certain extent, social networks, as social fabric woven in the threads of daily life and expressing collective ideas, concepts, convictions, aspirations, and aims, end up becoming tools for active citizenship and participatory democracy.

The Foundation of the Ibero-American Research Network in Education Policies (RIAIPE)

In 2006, a group of nine research centeres, gathering social scientists (and activists) from different fields – education, sociology, anthropology, political science, economics – and from different countries (Argentina, Brazil, Spain, Mexico, Paraguay, and Portugal), submitted a proposal for an Ibero-American Research Network in Education Policies (RIAIPE) to the Programme Science and Technology for Development (CYTED), established within the scope of the Organization of Ibero-American States (OEI). Approved for funding for a four-year period, the Network started its work in early 2007.

In this first stage, the key goal of the RIAIPE network was to coordinate research carried out in the field of education policy analysis by the integrating teams. It aimed to establish a theoretical and analytical framework which would enable it to map and analyze public policies on education in the last decades – those carried out by the governments as well as the proposals and projects put forward by the most influential globalizing agencies or by social movements and local administrations. The Network defined as its goal to reinforce (and coordinate) research carried out about the impacts of globalization on education public policies, particularly in the fields of inclusion and equity, in the countries of the Ibero-American space to which the integrating teams belonged. Taking this mapping as reference, we aimed to develop a set of indicators favoring the dimensions of inclusion and equity in public policies, which could be set forth as alternative to the hegemonic parameters developed within the scope of such organizations as the World Bank (WB) or the Organization for Economic Cooperation and Development (OECD), used ad nauseum in reports, exams, and comparative statistical surveys, and which have become an influential regulation tool of public policies.

The Network fostered the creation of strong scientific and academic cooperation bonds, as well as the development of new projects, both bilateral and multilateral. The most important result of the work done was undoubtedly the transfer of knowledge among teams, regarding epistemologies, methodologies, and practices. This mutual learning gave way to advanced training and researchers' participation in international conferences and seminars, as well as teams' publications and consequently their impact in their respective scientific communities.

Once the connection to the CYTED program came to an end, it was necessary to apply to other sources of funding. The EC-Alfa program, whose main goal was to foster cooperation among Higher Education Institutions (HEIs) of the European Union (EU) and LA, was the alternative sought. It enabled the Network to expand and develop a stronger intervention project, which could contribute to improve the quality, relevance, and democratization of Higher Education in LA, as well as the process of regional integration in LA, by promoting the move toward the creation of a common area of Higher Education in the region and by fostering its synergies with the EU's university system.

Inequality and exclusion were understood as two leading features in the HEIs of LA (and, to a different degree, increasingly in the EU itself). When approaching the struggle against these two overriding elements, we favored a solution made to emerge from within the higher education institutions themselves, e. g. inclusion policies in the HEIs, governance systems, relevance of university programs, within the framework of their respective national education systems, in a regional and world context where Education and Science constitute two strongly globalized fields. We started out by acknowledging the need to maintain a high level of cooperation among the HEIs participating in the network, by designing and implementing a common agenda for equity and policies and norms adjusted to each context. This enabled the participating teams to pinpoint decisive causes and factors in the existing situation and submit (as well as carry out) proposals aiming to overcome processes and mechanisms which exclude entire populations (indigenous ethnic groups, afro-descendants, the poor, disabled people, minorities) from attending (and succeeding in) higher education.

In this context, the action proposed within the scope of the RIAIPE3 project was adapted to the goal of the Alfa III Program: to reform and modernize Higher Education systems and institutions in the beneficiary countries, by paying special attention to the more disadvantaged and vulnerable groups and the poorest countries in the region. The general goal of the project involved the substantial increase of equity and, therefore qualitative improvement and social relevance of the HEIs. Its specific objective was to develop an Inter-University Framework Programme (IFP) which would promote the structural transformation of HEIs through intervention models that improved the relevance of university role toward a balanced social development, while fostering equity and social cohesion.

The rationale of the above actions is directed toward the strengthening and consolidation (empowerment) of universities as social agents and the reinforcement of their structures, their potential, and their coordination to define criteria of high social impact, while aiming to set themselves as reference to other networks, institutions, and social agents.

The RIAIPE3 Model

The Network and the Project

The *Inter-University Framework Program for Equity and Social Cohesion Policies in Higher Education* project (2010–2013) was funded by the ALFA III, EU-program, which was a cooperation program between the EU and LA, promoting higher education as a means of social and economic development and the struggle against social inequality. The ALFA Program (2007–2013) aimed at the modernization of Higher Education in LA and to act as a platform to promote sustainable and equitable development in the region. It intended to ensure a process of ownership by the Latin American countries through the creation of networks and synergies between universities in LA and Europe. In this regard, ALFA III was in line with the EU-Latin American's Common Higher Education Area objective, recognized as a strategic element for strengthening bilateral and multilateral relations between the two regions, where higher education institutions play a leading role in the process of improving the quality of national education systems which in turn enables socio-economic development (https://ec.europa.eu/europeaid/where/latin-america/regional cooperation/alfa/index_en.htm_en).

This project encompassed 22 teams from LA higher education institutions and 8 from European ones (from Portugal, Spain, Brazil, Mexico, Argentina, Paraguay, Chile, Uruguay, Peru, Costa Rica, Honduras, Guatemala, Salvador, Netherlands, UK, France, and Italy)[1] and interdisciplinary researchers, with a predominance of those placed in Education faculties. The project was coordinated by a Portuguese university, *Universidade Lusófona de Humanidades e Tecnologias*, scientifically represented by António Teodoro, general coordinator, and Manuela Guilherme, scientific co-coordinator. The work plan encompassed transversal work according to geographical Latin American sub-regions, namely field analysis reports and respective plans of action implementation. Simultaneously, conceptual clarification was undertaken through four thematic Committees, namely on "Gender", "Social Relevance and Citizenship", "Equity and Social Equity" and, finally, "Governance, Democracy, and Citizenship", all focusing on Higher Education institutions' governance, internationalization, research and extension activities with a strong concern for the vulnerable groups' inaccessibility to higher education and dropping-out rates. Each committee was chaired by one Latin-American and one European coordinator and produced a great number of different publications related to their research topics. Besides many workshops and debate *fora* held at Latin American universities, the project regularly organized, throughout the 3 years, six 2-day general meetings, all preceded by a public 2-day conference, which were held at Lisbon (Portugal), La Paz (Bolivia), La Habana (Cuba), Guadalajara (Mexico), and Salvador da Bahia (Brazil). During the last meeting in S. Paulo (Brazil), project coordinators, evaluators, committee coordinators as well as every team coordinator were given the opportunity to video record their evaluation statements either individually or in groups.

The Participants' Statements: Communication and Collaboration

The analysis of the final video recorded statements (list below), still accessible online, provides a wealth of interesting critical thoughts not only about this project in particular but also about the limits, risks and great potential that international and intercontinental research networks can offer. Despite the large size of this particular RIAIPE3 network, which required careful coordination, it was possible to balance both the freedom, the participant researchers needed to breath, create, and attend contextual needs, and a strict organization of deadlines, deliverables, and outcomes. On the whole, it is evident that all participants found this an enriching experience, generating such high expectations that these justified some frustration about the achievements that remained still to be accomplished by the end of the project. However, they were also hopeful and determined to continue the path they had just started mainly with regard to institutional change and community service. The project coordinators could particularly appreciate that, on the whole, the participants affirmed that the communication and interaction among all researchers, both Latin American and Europeans, was horizontal and mutually enriching. On the one hand, one of the South American researchers, from Chile, mentioned that he felt that, during this project, Latin American colleagues were addressed as "valid interlocutors", they were not "told to" by European colleagues and that the level in which this was so had been somewhat unexpected to him. Unexpected, this was for European researchers too, which makes evident that there are communication and interaction failures of which both parties need to be made equally aware. According to him, this cooperative standing had promoted the establishment of epistemological "bridges". On the other hand, another researcher from Central America said that he had not interacted much with the European colleagues, but that he had profited from the materials they had posted on the platform. We can therefore conclude that the perception of accomplishment in epistemological and social communication and collaboration varied according to individual researchers, to collective tasks at different stages and to communication media.

It is important to note that all variables, namely communication and interaction between participants as well as institutional impact, were not uniform, given the large number of participants and the reduced, although regular, opportunities for face-to-face communication and interaction as well as the contextual diversity, both regional and institutional. The degree of acquaintance between participants started at very diverse levels, since some had already collaborated in previous projects or even met in previous events, and also ended at different levels due to the groups dynamics. It is also noteworthy the fact that different languages were at play and participants also had different levels of proficiency at different languages and, therefore, both linguistic proficiency and linguistic audacity somehow also prompted social relations and professional collaboration. Portuguese and Spanish were dominant and English was the language used between anyone proficient enough to communicate with northern Europeans. However, the use of English was balanced with that of Portuguese or Spanish with regard to publications whereas it was predominant in terms of reports to be delivered to the European Commission. In addition, it is also significant to refer, as most of the recorded statements mention, the intricacies of communication between different varieties of Spanish and Portuguese, within each language and between both, which participants highlight as most enriching, mainly in conceptual discussions and writings.

Contextual Diversity among Different Institutions of Higher Education

All Latin American participants also explain, in their video recorded statements, the different stages at which equity issues were approached in their own HEIs, e.g. with regard to gender and LGBT and economically vulnerable communities, among them indigenous communities, namely as far as university students access and permanence were concerned. For example, with regard to the teaching staff they reported that gender discrimination in higher education careers was seldom in the agenda. On the whole, although at least some of these problems had already been taken into account, they had not yet been made sufficiently visible or tackled systematically, with institutional regulations enforced. The project, according to their statements, allowed them to introduce these debates with the legitimacy and pressure given by an international network. Such statements give evidence of the power of higher education internationalization and research transnationalization for shaking local power games and cultural traditions. Another example was that adequacy of semantics to gender equity had never been raised. Furthermore, critical discussions about discrimination and lack of equity were widely not even considered appropriate to be carried out in higher education professional environments and mainly by the academic community, especially when those targeted the academy community itself. Some participant researchers promoted debates for the whole academic community, in their own institutions, about the integration of indigenous students in their institutions and departments and those had felt empowered to share their experiences and even to talk reflexively about their own cultural traditions. One of the participating research teams, in Bolivia, had managed to establish a formal collaboration agreement with one indigenous community, in order to proceed with debates and activities relevant to them and their representatives at the university.

Participant researchers reported that they had raised those issues of equity and social cohesion referred above, within their institutions, through queries, workshops and, in some cases, managing curricular changes and additions. They also referred to other colleagues from other national and regional universities joining the debate and promoting joint initiatives. The project impact, however, depended on distinct factors, namely the size of each institution, larger universities, such as those in Argentina, required more bureaucracy and this made it more difficult to reach the centeres of power, while in smaller institutions of higher education it was more feasible even to involve the institutional authorities in the project initiatives and "to interrupt inertia", that is, to manage to put equity issues, those more relevant to their specific context, in the institutional agenda. By that time, with the start of the second decade of the 21st century, all participant institutions were prone to increase their internationalization, a feature that had just started to be very valuable for institutional ranking and social image, therefore, such a large network project, moreover with the support of the European Commission and the background of the European Higher Education Area, was an inviting coincidence.

Local and Regional Impact, Interculturality, and Internationalization

Many of the participating HEIs were only starting to address equity problems among their students which had caused them to lag behind in their academic progression and eventually to drop-out, as well among the vulnerable groups who did not even have access to higher education. In Guatemala, for example, university is, on the whole, inaccessible to rural populations and this restriction is replicated in most participant countries which makes it a general lack of equity that cannot be disregarded. This was also a problem that was helpful to share since some of the participant institutions were using virtual platforms not only to attract students from rural areas but also to involve them in activities through this media. Many of the Latin American nations where the participating institutions were placed, were at the time undergoing democratic political developments which have unfortunately been currently reversed, such as Brazil, Argentina, and others. Political instability in LA, alternating between socialist and capitalist governments, not to mention ingrained corruption on both sides, has prevented policies which might both protect and stimulate vulnerable groups from reversing deep-rooted discriminatory practices. Contextual political situation strongly determines the success or failure of a project with objectives focusing on equity and social cohesion. Brazil is an unfortunate example. This project had just started when the Dilma Rousseff government approved a law enforcing quotas for economically disadvantaged candidates to federal universities as well as for Afro-Brazilian, indigenous, and disabled students. The Lula government had expanded the federal system of higher education to poor areas and to working students, also founding popular, rural, border, and international universities. Now, all these policies are being reversed.

Most participant researchers, e.g. from Chile, pointed out the fundamental role this project/network had played in crossing geographical and political barriers, breaking off parochialism and opening up fields for collaborative knowledge, which another colleague from Argentina called "living interculturality". The relevance of virtual communication, once supported by face-to-face meetings, was also highlighted. However, it needs to be reminded that it was common understanding, that all this was optimized by a work plan which included simultaneously actual practice, in the form of field analyses and plans of action, with conceptual clarification, not only through lively debates but also through hands-on publication outcomes which meant that peers had to eventually arrive at solid and grounded conclusions or, at least, question-posing. Some participants also noted that not all peers had had previous academic international experience nor were them all prepared for it, at different stages, when the project started, however, the project dynamics encouraged them to leave their comfort zone at each one's particular pace. Interestingly enough, the size of the project enabled some to go through this individual internationalization process almost incognito, both in the project and in their institutions, while nevertheless their intercultural development progressed throughout the different specific collaborations at various levels and in different groups, both within the geographical regions and across the project thematic fields. Different dynamics allowed them the freedom to breathe whenever needed and move at different pace toward the fulfilment of various tasks. The result was an enormous amount of productivity, tempered with creativity and autonomy, at different levels. Team coordinators also mentioned the advantages that young researchers, both postgraduate students and research assistants, had taken from the project dynamics and the network being expanded and consolidated which gave them prolific opportunities for starting helpful collaborations and study visits that allowed them to further and improve their research projects.

Different types of HEIs also enhanced different types of approaches to the themes of the project. A colleague from Brazil, working in a communitarian university particularly concerned with issues of social inclusion, managed to bring the scientific study of social problems in this particular community to the postgraduate programs and professional development programs, which gained more legitimacy due to the project/network international scope and prestige. In addition, an Argentinian colleague, from a different type of university, voiced exactly the same conclusion, that is, the project work plan, by linking theory and practice, strengthened the relationship between the teaching, research and extension (connection with the communities to which the students or students-to-be have to proceed), which was though already foreseen by Latin American universities' academic life programs.

Latin America and European Higher Education between Mirrors

This project put Latin American and European universities, as well as their teaching and research staff, between mirrors (Teodoro and Guilherme 2014). Both Latin American and European participants concluded that this cooperation had been very fruitful and that they had learned much from each other. An Argentinian colleague concluded that the role of the European partners was most enriching but different, that is, since the focus was only on Latin American HEIs, they had provided other views on Latin American problems as well as a critical understanding of the European Higher Education Area which had contributed to develop a general perception of risks and possibilities in the implementation of the Latin American Higher Education Area. Although the European partners were in a much more reduced number and were not the focus of the project, their participation was neither less nor more predominant than their Latin American peers', since they intensively collaborated in all the project work and integrated the general structure of the project in parity with their Latin American peers. With regard to the two project Evaluators, one was Portuguese and the other was of Argentinian origin but a Professor the University of California Los Angeles. The former stated, in his video recorded statement, that the attitude of the European partners had not been one of helping but of cooperation and that he could acknowledge how much Europe has to learn from their Latin American partners as far as higher education theory and practice are concerned. In fact, this sense of mutual learning was very present in the minds of every participant and both the Latin American and the European researchers mixed and collaborated indiscriminately, proving how this plurilingual, intercultural, and intercontinental cooperation can be extremely productive for all sides. The coordinator of one European team, a Professor in the Netherlands with much experience of institutional assessment of higher education abroad, was particularly enthusiastic in his video recorded statement about this collaboration. He stated that Europe has much to learn from the ways how LA HEIs approach issues of educational, cultural and social development. He confessed: – "I probably learned more from them than they learned from me". And he added: "They work more on democracy and on social cohesion than many countries in Europe do". We dare say that this happens because most European institutions take it for granted that they have reached the top of a linear move to progress, but they should have been made to confront themselves more directly with the reality of vulnerable groups of new immigrants and refugees as well as the poverty pockets among the unemployed youngsters and, it should not be forgotten, the aged people hit by reduced pensions and defunded public services.

Both parties, European and Latin American researchers, followed different paradigms, but they were nevertheless capable to build a common framework and to start new research lines, that is, it was possible to unite research teams from distinct contexts toward common objectives. It was also common understanding that most of the links established between individual researchers and institutional agreements were to remain. According to some of the committee coordinators, from Spain and Honduras amongst others, there was not a closure in the work of each committee, on the contrary, the findings were put to discussion by the whole project team members. At the same time, the indicators found in relation to the project theme were very similar throughout the many Latin American countries in the project, which strengthened the exchange of experiences with regard to institutional plans of action promoted by the project. Therefore, one of the coordinators mentioned that a few transversal lines of research had emerged from the project development and these could live beyond the project. According to another committee coordinator, the main legacy of this project was precisely a networking organization model that combines a critical vision, which leads into intervention and transformation, with social and political commitment. A team coordinator, whose institution has carried out several projects in Africa, suggested that this model could be replicated in another continent, Africa for example. A Brazilian Professor, a companion to Paulo Freire and co-founder of the Paulo Freire Institute, affirmed that the project participants had no idea of the dimension of the impact that this ambitious project/network altogether had had in LA as a whole. Most participants enhanced the increase and improvement of South-South and South-North dialogue and the corresponding relationships established through this project/network collaboration. The statements above highlight that such research collaborations, across languages, ethnic and institutional cultures, epistemologies, and traditions, are not only possible, they can be fruitful and inspirational.

Conclusion

This chapter's main objective was to give the voice back to the RIAIPE3 participant researchers, let them reflect upon and evaluate the scientific organization of such a large network and the impact that it had had both individually, as individual researchers participating in an international and intercontinental network, and collectively, both at institutional and local team levels, and to thematic international and intercontinental committees, having in mind the particular aims of the RIAIPE3 project. Eight years have already elapsed since its completion, and we can gather from the above considerations that there was hard work involved, which generated a strong and dynamic network, only the beginning of a wide and bold regional impact together with an impressive production of outcomes, such as publications, workshops, and events, where scholars from different universities, countries and regions converged. According to the main actors, three years were not enough to achieve the promised structural institutional or societal relevant changes. It was only enough to oil and start the machine ...

We are then left with the question: Has such collaboration dynamics survived and lasted? Which were the expectations of the funding agency? The Alpha program has been cancelled but other opportunities have replaced it. The sad news is that the network, as a whole, has been left lingering, that is, it has been put to wait, while some members have reorganized and regrouped, adding new elements and losing others, in order to respond to the EU (and LA) funding relying on "call-systems". By way of conclusion, we dare say that it should be valuable to accurately examine and to reflect upon the gains and losses of such "call-systems" that call for short period projects and networks, imposing pauses and breaks and requiring constant rail and staff change. In this case, the steam, which had demanded a lot of energy during this 3-year period, abruptly faded away. The 2013 "Communication from the Commission to the European Parliament, the Council, the European Economic and Social Committee and the Committee of the Regions" about "European higher education in the world", though, explicitly stated: "But internationalisation requires cooperation, with new higher education hubs on other continents". It seems then pertinent to ask what we are supposed to do when we have just created one? Move on and create another one? What is, indeed, the ultimate goal?

Video Recorded Testimonies

Memórias RIAIPE3 | Statement - António Teodoro
https://www.youtube.com/watch?v=eQZW-WnP6hk
Memórias RIAIPE3 | Manuela Guilherme
https://www.youtube.com/watch?v=fUC0sf0EoaY
Memórias RIAIPE3 | Statement - António Magalhães
https://www.youtube.com/watch?v=HQpG-NiKn2w
Memórias RIAIPE3 | Statement - J. E. Romão
https://www.youtube.com/watch?v=bwDAFWHAyqE
Memórias RIAIPE3 | Statement - Boris Tristá
https://www.youtube.com/watch?v=rFXqtKYDk_Q
Memórias RIAIPE3 | Statement - Silvia Llomovatte
https://www.youtube.com/watch?v=nAWfV5HJQQw
Memórias RIAIPE3 | Statement - Paulo Peixoto
https://www.youtube.com/watch?v=7piLqcc4BVg
Memórias RIAIPE3 | Statement - Wiel Weugelers
https://www.youtube.com/watch?v=yVT6sUzymvs
Memórias RIAIPE3 | Statement - Jean-Claude Regnier
https://www.youtube.com/watch?v=sWEfl-rQewo
Ângela Santamaria | Universidad, Colombia
https://www.youtube.com/watch?v=jD9ny83-i_I
Memórias RIAIPE3 | Comité 1 - Armando Alcantara, Sandra Montané, Alejandrina Mata
https://www.youtube.com/watch?v=bG-r1HW_Ndc
Memórias RIAIPE3 | Comité 2 - José Beltrán, Claudia Iriarte
https://www.youtube.com/watch?v=FaB8sgJx4vU
Memórias RIAIPE3 | Grupo 1 - Clara Almada, Fernando Cajas, Robinson Tenório
https://www.youtube.com/watch?v=Q08H1bo5MUc
Memórias RIAIPE3 | Grupo 2 - Ana Donini, Mª Graça Bollmann
https://www.youtube.com/watch?v=A9XpezMxlKM
Memórias RIAIPE3 | Grupo 3 - Carmen Velezmoro, Javier Merlo, Christian Mendizábal

https://www.youtube.com/watch?v=rNhkuuRUh7o
Memórias RIAIPE3 | Grupo 4 - Godofredo Aguillón, Liliana Olmos, Carlos Guazmayan
https://www.youtube.com/watch?v=YQhh1MHrB8A

Note

1. The RIAIPE3 network included the following higher Education institutions both from Europe and Latin America: Universidade Lusófona de Humanidades e Tecnologias (coordenador) and Centro de Estudos Sociais (Universidade de Coimbra), Portugal; Universitat de Barcelona and Universitat de Valencia (Espanha); Université Lumière Lyon 2 (França); Universiteit Voor Humanistiek (Holanda); Università degli Studi della Tuscia, Itália; University of Brighton (United Kingdom); Universidad de Buenos Aires, Universidad Nacional de la Plata/Instituto Paulo Freire, Universidad Nacional de San Martín e Universidad Nacional de Tres de Febrero, Argentina; Universidad Loyola de Bolivia, Bolivia; Universidade Federal da Bahia, Universidade Nove de Julho e Universidade do Sul de Santa Catarina, Brasil; Universidad de Ciencias de la Informática, Chile; Universidad de Nariño e Universidad del Rosário, Colômbia; Universidad de Costa Rica, Costa Rica; Universidad de La Habana, Cuba; Universidad de El Salvador, El Salvador; Centro Universitario de Occidente (Universidad de San Carlos de Guatemala), Guatemala; Universidad Nacional Autónoma de Honduras, Honduras; Universidad Autónoma de México e Universidad de Guadalajara, México; Universidad Autónoma de Asunción e Universidad Nacional de Asunción, Paraguai; Universidad Nacional Agraria La Molina, Perú; Universidad de la Republica, Uruguai. Participaram, como associados, a Università di Bologna, Itália, a Universidade Federal da Paraíba, Brasil, as well as the Organisation of the Iberoamerican States.

References

Altbach, P. G. (2013). *The International Imperative on Higher Education.* Rotterdam: Sense

Arendt, H. (1958). *The Human Condition.* Chicago: Chicago University Press. 2nd Ed, with a Preface of Margaret Canovan, Chicago & London, 1998

Ball, S. (2016). Following policy: Networks, network ethnography and education policy mobilities. Journal of Education Policy, 31: 5, 549–566

Banerjee, P., Nigam, A. and Paney, R. (2016). The work of theory: Thinking across traditions. Economic and Political Weekly, 51: 37, 42–50

Canclini, N. G. (2003). *A Globalização Imaginada.* S. Paulo: Iluminuras

Castells, M. (1996). *The Rise of Network Society.* Oxford, UK: Blackwell Publishing

Castro-Goméz, S. and Grosfoguel, R. (2007). *El giro decolonial: Reflexiones para una diversidad epistémica más allá del capitalismo global.* Bogotá: Siglo del Hombre

De Wit, H. (2011). *Trends, Issues and Challenges in Internationalisation of Higher Education.* Amsterdam: Centre for Applied Research on Economics & Management, School of Economics Management of the Hogeschool van Amsterdam

Freire, P. (1985). *The Politics of Education: Culture, Power, and Liberation.* Westport (CT): Greenwood Publishing Group

Giddens, A. (1994). *Beyond Left and Right. The Future of Radical Politics.* Cambridge, UK: Polity Press

Guilherme, M. and Santamaria, A. (2015) (eds.) Ventos do Sul: Modelos e epistemologias interculturais emergentes na educação superior na América Latina. *Revista Lusófona de Educação*, 31: 31

Guilherme, M. and Lourenço, F. (2015) Da praxis de-colonial e intercultural no ensino superior indígena: Andante ma non troppo. In M. Guilherme and A. Santamaria (eds.) Ventos do Sul: Modelos e epistemologias interculturais emergentes na educação superior na América Latina. *Revista Lusófona de Educação*, special issue, 31: 31, 179–197

Guilherme, M. and Dietz, G. (2017) (eds.) Winds of the south: Intercultural university models for the 21st century. *Arts and Humanities in Higher Education*, Special issue, 16: 1

Habermas, J. (1962/1989). *The Structural Transformation of the Public Sphere: An Inquiry into a Category of Bourgeois Society.* Polity: Cambridge

Hudzik, J. K. (2015). *Comprehensive Internationalization: Institutional Pathways to Success.* New York, NY: Routledge

Kehm, B. M. and Teichler, U. (2007). Research on internationalisation in higher education. *Journal of Studies in International Education*, 11: 3/4, 260–273

Leite, D. and Pinho I. (2017). *Evaluating Collaboration Networks in Higher Education Research.* Cham, Switzerland: Springer

Mignolo, W. (2000). *Local Histories/Global Designs: Coloniality, Subaltern Knowledges, and Border Thinking.* Princeton, NJ: Princeton University Press

Mignolo, W. (2007). El pensamiento decolonial. Desprendimento y apertura. Un manifiesto. In S. Castro-Gómez & R. Grosfoguel (eds) *El giro decolonial: Reflexiones para una diversidad epistémica más allá del capitalismo global* (pp. 25–46). Bogotá, Colombia: Siglo del Hombre Editores

Mignolo, W. and Walsh, C. (2018). *On Decoloniality: Concepts, Analytics, Praxis.* Durham: Duke University Press

Red Iberoamericana de Investigaciones en Políticas Educativas. (2012). *La Educación Superior en el Mercosur: Argentina, Brasil, Paraguay y Uruguay hoy.* Buenos Aires: Editorial Biblos

Rumbley, L. E. and Altbach, P. G. (2016). The local and the global in higher education internationalization. In E. Jones, R. Coelen, J. Beelen and H. De Wit (eds.) *Global and Local Internationalization* (pp. 7–13). Rotterdam: Sense Publishers

Santos, B. S. (2007). Beyond Abyssal Thinking: From global lines to ecologies of knowledges. *Review*, 30: 1, 45–89

Santos, B. S. (2014). *Epistemologies of the South.* Boulder: Paradigm

Santos, B. S. (2017a). *Decolonising the University: The Challenge of Deep Cognitive Justice.* Cambridge: Cambridge Scholars Pub

Santos, B. S. (2017b). A new vision of Europe: Learning from the South. In G. K. Bhambra & J. Narayan (eds.) *European Cosmopolitanism: Colonial Histories and Postcolonial Societies* (pp. 172–184). New York, NY: Routledge

Spivak, G. C. (1999). *A Critique of Postcolonial Reason.* Cambridge, MA: Harvard University Press

Teodoro, A., Mendizábal Cabrera, Lourenço, F. and Villegas Roca, M. (eds.) (2013) *Interculturalidad y Educación Superior: Desafíos de la diversidade para un cambio educativo*. Buenos Aires: Editorial Biblos

Teodoro, A. and Guilherme, M. (eds.) (2014) *European and Latin American Higher Education Between Mirrors: Conceptual Frameworks and Policies of Equity and Social Cohesion*. Rotterdam: Sense

Teodoro, A., Torres, C. A. and Romão, J. E. (2014). Institutional Networks in Latin America. Building New Paths in Academic Cooperation. In A. Teodoro and M. Guilherme (Ed.). *European and Latin American Higher Education Between Mirrors. Conceptual Frameworks and Policies of Equity and Social Cohesion* (pp. 75–90). Rotterdam: Sense Publishers

Urry, J. (2005). The complexities of the global. *Theory, Culture & Society*, 22: 5, 235–254

Visvanathan, S. (2007). Between cosmology and system: The heuristics of a dissenting imagination. In B. Sousa Santos (ed.) *Another Knowledge Is Possible* (pp. 182–218). London: Verso

3 The impacts of institutional flexibility, incompleteness, and rigidity on the quality of democracy

The building up of historical-oriented concepts[1]

Marta Maria Assumpção-Rodrigues[2] *& José Veríssimo Romão Netto*[3]

Introduction

Do flexible institutions enhance democracy? What are the democratic implications of flexible and incomplete governance institutions? What role do this new institutional design play in countries where democracy is not a tradition? What is the role of flexibility and incompleteness in contexts where the institutional design does not enhance the belief in the legality of enacted rules (Weber 1978)?

In order to answer these questions, this chapter proposes that in political science, in general, as in the field of public policy, in particular, concepts such as "flexibility" or "incompleteness" may not necessarily produce democratically beneficial effects – but rather "rigidity". According to Goodin, for instance, political institutions need to be flexible, but not "brittle"; that is to say, they should just "be able to *adapt* to new circumstances, without being *destroyed* by them" (1996: 40). Lowndes (2014), in turn, has considered incompleteness as "a design value in itself", since it is

> "neither a good nor a bad thing (in itself), but rather an intrinsic property of institutional design. Signifying stability (i.e, a compromise that 'works', at least for a while) *or* instability (a punctuation within path dependence), the notion of incompleteness also relates to 'either less or more,' i.e., it relates 'to the *deficit* or the *asset* concepts'".

Combining these approaches with the notion of flexibility, defined, in institutional terms, as "a set of good, or good enough design" (Lowndes and Roberts 2013), and given the realities of institutional embeddedness of mature democracies (including, of course, the case of the United Kingdom), incompleteness, together with flexibility, is a notion that may not only facilitate the enhancement of institutional innovation and

DOI: 10.4324/9781003225812-4

learning but may also lead to institutional alterations, where appropriate, and to the improvement of democratic quality.

Looking at some Latin American countries, however, it is possible to approach the questions presented at the outset of this chapter stating that institutional flexibility and/or incompleteness, in these cases, may enhance either pernicious aspects of *political life* – such as low-intensity citizenship (O'Donnell 2005) – or detrimental effects of how people *do politics* – such as clientelism (DaMatta 1995; Nunes 1984). After all, the democratic deficit, in some of these cases, may lead to a situation where rules simply do not "stick" for all but only for a small audience. In any case, in contexts where the liberal and republican components of a polyarchy seem to be absent, the "embedded autonomy" of the state – meaning that its relation with society is mediated by a committed, meritocratic, and efficient bureaucracy (Evans 1995) – gives space for "institutional insulation" (Nunes 1984) which means ... Consequently, the democratic quality of such political institutions may fade away.

Taking these considerations into account, it is worth mentioning that this chapter attempts to challenge the assumption that there is a strong (and positive) correlation between institutional flexibility/incompleteness and the emergence of new channels of political participation, as well the engagement of civil society in policy processes (formulation, implementation, and evaluation of public policies). Besides, it argues that greater attention must be paid to the *relationship* between institutional designs and the quality of democracy, which relies, on the one hand, upon the existence of policy continuities (or discontinuities), in terms of the reach of the state's legality in several regions of a country (O'Donnell 2004), and refers, on the other hand, to the extent to which political life and institutional performance of public policies coincides with citizens' aspirations (Vargas Cullell 2004).

Quality of Democracy, Citizenship, vs. Institutional Rigidity

Thus, as a result of the research agenda shared by the Centre for Public Policy Research at the University of São Paulo with the Institute of Local Government Studies at the University of Birmingham, from 2014 to 2016, our investigation departed from the proposition that, considering the "south/north" perspective, the idea that specific institutional arrangements can operate in diverse local environments must be challenged, since, implying different world visions, these arrangements remain geographically situated. All this take us back to the necessity of looking at the *relationship* between the institutional design (in terms of flexibility/incompleteness/rigidity), and the quality of democracy approaches.

The initial focus of our investigation was on the issue: Do flexible institutions enhance democracy? However, this issue was rapidly questioned by two initial and provocative documents. One, entitled "Conceptualising Incomplete Institutional Design", was written by Vivien Lowndes (2014) of the Institute of Local Government Studies – INLOGOV, and the other, "Conceptualising Quality of Democracy", was produced by Marta M. Assumpção-Rodrigues (2015) of the Centre for Public Policy Research – NUPPs/USP. The latter work followed O'Donnell when he defined "institutions" as

> "regularized patterns of interaction that are known, practiced, and regularly accepted (if not necessarily normatively approved) by social agents who expect to continue interacting under the rules and norms formally or informally embodied in those patterns. Sometimes, but not necessarily, institutions become formal organizations: they materialize in buildings, seals, rituals, and persons in roles that authorize them to 'speak for' the organization"
>
> (1994: 57)

As a consequence, the debates about these initial reflections were rapidly followed by a derived theoretical concern (Skelcher and Romão Netto 2016), which reframed our questions, since our research was interested in exploring *intentional* incompleteness:

What are the democratic implications of new, flexible governance institutions? What are the democratic consequences of institutional flexibility/incompleteness, and rigidity? How can governments and citizens design innovative institutions that will improve the quality of democracy in public policy-making? Or should greater power be given to political elites or state actors (like civil servants, state professionals etc)?

Theoretical Background

The study of democratic regimes by scholars of comparative politics experienced several interesting turns in the last decades. From comings and goings of democratic political regimes, whose elastic texture seems always to be open to new developments, we have learned a lot. Thus, it might be useful to start this section with the idea that democracy is an incomplete political phenomenon by nature.

To illustrate this point, it is worth to begin by emphasizing that, in the 1970s and 1980s, for example, political developments in Southern European (O'Donnell et al. 1986a), Latin American (O'Donnell et al. 1986b), and Central and Eastern European (O'Donnell et al. 1986c) countries prompted a concern of several scholars with regime change, in general, and with transitions to democracy, in particular.

Nevertheless, as the third wave of democratization (Huntington 1991) spread to other parts of the world, scholars shifted their focus to democratic consolidation (Mainwaring and Hagopian 2005; Mainwaring

1999; Diamond and Plattner 1997) or, as Plattner (2005) has stated: "from the ways in which democratic regimes come into being to the ways in which they can be rendered stable and secure". Today a double concern seems to prevail in relation to the study of democratic regimes.

Furthermore, on the one hand, the regression to authoritarianism after processes of democratization in some countries – of which Russia is perhaps the negative best example – and, on the other hand, the emergence of semi-democracies or hybrid regimes (Skelcher 2012), provoked a new interest in the study of authoritarian legacies in the context of the new democracies, competitive authoritarianism and even new dictatorships (Levistky and Way 2010). In fact, the study of the dynamics of democracy has never really lost its appeal, quite on the contrary, it is more and more a central concern of the intellectual debate on contemporary political science.

With the end of the Cold War and the expansion of liberal democracy worldwide, however, academics and practitioners have begun to reflect upon a new concern: no longer the "transition" to democracy or the "consolidation" of the democratic regime, but rather the "quality of democracy". As Adam Przeworski has recently written on this matter, "for all of us who had followed liberalization, transition and consolidation, we have discovered that there is still something to improve: democracy" (Przeworski 2010). Nevertheless, the term quality of democracy has become a buzzword in political science – especially to scholars from a more traditional democratic background.

Quality of democracy has been associated with the themes that are useful not only to suggest main characteristics of contemporary democratic regimes around the world but they are also particularly useful to shed some light on current dilemmas that new democracies have faced – like the nature of the democratic regime (Diamond and Morlino 2005; Vargas Cullell 2004), of governance (Bovaird and Loeffler 2015), and of the policy-making process – including the institutional design (Hallerberg et al 2009; Stein and Tommasi 2005; O'Donnell 2004).

Moreover, if old forms of instability, leading to regime breakdown and transitions (Linz and Stepan 1996; Przeworski 1986) have ceased to be a major concern to most of the academic world – and particularly in Latin America – new types of instability, in the form of failed presidencies or failed political coalitions (Llanos and Marsteintredt 2010; Pérez-Liñán 2007), episodic constitutional crisis (Boniface 2002; Brinks 2010), fragmentation and dysfunctionality of the party system (Mainwaring and Torcal 2006; Mainwaring 1999), and increasing political protest often followed by political intolerance (Mainwaring and Pérez-Liñán 2013), abound.

In spite of the fact that these new forms of instability – which Lowndes (2014) has referred to as "a punctuation within path dependence"[4] – do not represent immediate risk of democratic regime breakdown, they can certainly leave deep scars in political relations between electoral winners and losers, citizens and authorities, government and opposition, and pose clear threats to the regime legitimacy. At the same time, and according to a more optimistic view, several scholars have stated that the existing ineffectiveness of old and new democracies in responding to new and old demands – as expressed by the crisis of representation – is related to increased experimentalism with new forms of political participation and deliberation (Pogrebinski, 2014; Avritzer and Santos 2002; Avritzer 2000) – that Lowndes (2014) calls "a compromise that 'works' at least for a while". This new reality not only produces or reinforces a contrast in cognitive levels about the understanding of what democracy is, but it is also fundamental to the support for and the legitimization of the regime. All these dimensions are essential components of existing democracies.

Hence, understanding if and when these new alternatives are a way of capturing "new recipes" for democratic governance is also important (Avritzer 2017; Diamond and Morlino 2005). At the end of the day, legacies affect the degree of respect for the rule of law, of incomplete levels and forms of political participation, and of a higher or lower level of trust in institutions.

From the perspective of more traditional democracies, these dilemmas have also been exacerbated by the emergency of new and sometimes *flexible* modes and channels for citizens' direct participation, political protest and political tolerance. The debate over new complex arrangements for governance that have been brought about, and their challenges for the classical representative democracy and its core institutions (Feldman and Khademian 2007; Vangen and Huxham 2003), has engaged politicians, civil society organizations, the high bureaucracy of the state, as well as academic researchers. Moreover, at the heart of this debate is the question of how government and citizens can design innovative institutions that will enhance democracy in public policy making, and also increase structures for transparency, responsiveness, and accountability (Filgueiras 2018; Olsen 2013; Skelcher et al. 2005). Besides, whether all this occurs in conditions which includes political tolerance in facing political differences is also a crucial aspect of this new democratic picture.

The Rationale for a Brazil/UK Comparison

The rationale for a Brazil-UK comparison is that the two countries have also exhibited some similarities in the development of new forms of flexible public governance, despite different underlying democratic conditions within which these are located, and in which they will, thus, be impacted. For instance, in both countries, new institutions of governance have been developed to bridge the gap between state and civil society. They offer the potential to enhance mainstream democratic relationships by opening up new spaces at arm's length to elected politicians for the purpose of articulation, deliberation, and resolution of claims by citizens on the resources of the state and the form and operation of its public policies.

With the democratization process, Brazil may be regarded as a case in which the authoritarian heritage was followed by civic and administrative efforts to reshape democratic institutions and political culture (Romão Netto 2015; 2010; Bresser Pereira and Spink 2006). The 1988 Constitution, under the influence of the actions of NGOs and social movements that had been struggling against authoritarianism, allowed new kinds of popular organizations to become incorporated in the structure of the state, some of which were related to decision-making on public policy, and others to the process of monitoring how it is put into practice besides other instruments of direct social participation, such as plebiscites, referendums and legislative initiatives – as well as Public Policy Managing Councils (Health and Education etc) (Doimo and Assumpção-Rodrigues 2003), and Participative Budgeting (Gurza Lavalle 2011; Avritzer 2009; Coelho 2004).

The major innovation that accompanied the process of democratization of the 1980s was the expansion, on the part of the state, into institutionalized mechanisms for popular participation. It is worth recalling that from the 1988 Constitution onward Brazil becomes a federative republic. The purpose of these initiatives was to address the lack of governability (Skelcher and Romão Netto 2016).

Nevertheless, the relations between the Brazilian civil society with the political and administrative structures of government during the democratization period have been the object of a host of studies, to the point that the terms "*ungovernability*" and "*ingovernance*" (Aguilar Villanueva 2009; Melo 1995) were created in order to address the inability of the government in meeting social demands for participation – which resulted in generalized dissatisfaction on the part of population (Assumpção-Rodrigues 2018). Then, the answer to this problem involved both the mobilization of the institutional resources of the state (in both its political and technical dimensions), and the capacity for making political coalitions in order to maintain democracy. Needless to mention, however, that several attempts to establish mechanisms for communication with society, including the project for State Reform of the 1990s proposed by the Ministry of Federal Administration and State Reform of Cardoso administration (Skelcher and Romão Netto 2016), failed.

More recently, however, these coalitions that involve, on the one hand, the interpenetration of different spheres of activity – government, business, civil society, not-for-profit organizations –, as well as, on the other hand, structured interconnections through parastatal organizations - such as public-private partnerships, collaborative management, and policy networks – have demonstrated to be pernicious as they facilitate widespread corruption, clientelism, and other informalities that O'Donnell named "particularistic practices" (2010).

> These are particularistic practices, manifested not only in sheer corruption but also in nepotism, clientelism, discriminatory application of legal rules, abusive use of the perks of office, and others. [...]
>
> But even though in this sense the state as thus embodied is still "present", the state as a legal system has evaporated because of a perverse privatization, by which the public aspect of the state as law is "sold" by means of particularistic transactions.

In the UK, in turn, where there is a longer tradition of representative democracy and parliamentary government, legitimation crises have stimulated a strong political narrative that has promoted greater involvement of civil society (citizens and NGOs) in the formulation and delivery of public policy (Durose et al. 2009). In this case, such involvements are often located in "partnerships", actively developed by the Blair government from 1997 to 2008, and incentivized by a funding regime that required the creation of stakeholder partnerships in order to gain access to resources as well as a managerial ethos that valued involvement of citizens and other actors in public policy processes. Subsequently, there has also been the development of other single purpose of public bodies operating with semi-autonomy from elected politicians.

Overall, in the case of the UK, there has been substantial growth in quasi-governmental institutions at neighborhood, city, and regional levels across a wide range of public policy fields. These bodies shape policy, sometimes decide budget allocations, and oversee the delivery of public services (Sullivan and Skelcher 2002). In addition, there has also been a strong political interest in the development of public engagement in neighborhood governance (city and sub-national levels) involving, in addition, the forging of relationships between public service professionals and civic activists, and the overseeing of public policy delivery.

In summary, all these developments in Brazil and the UK are part of a global process of public service reform that has resulted in a great variety of institutional innovation for shaping, making, and delivering public policy. Such institutions, because they operate from elected politicians, and often are strongly influenced by non-state actors (citizens, professions, businesses, NGOs), typically have a greater degree of flexibility or incompleteness in their design than is the case with conventional ministries and departments of government – especially in countries that have a formal constitution and public law system, like Brazil. Consequently, questions of democratic quality, in terms of how democracy can be enhanced and the political power distributed or accessed, cannot be ignored. These institutions debate and decide the distribution of public welfare to the citizens or users within their jurisdiction. This is an inherently political process, and the effectiveness with which normal constitutional safeguards apply, thus, becomes an important matter for academic inquiry and policy debate.

Mixing up Historical Oriented Concepts

Firm answers to many important questions posed by our investigation beg for future comparative research. At this point, however, it is worth noting that our intellectual move was just to involve the "incomplete/flexible and/or rigid institutional approach" with that of the "quality of democracy", which were understood by policy researchers from different parts of the world as intertwined concepts.

In these terms, the very notion of quality of democracy brought about from these discussions referred to the extent to which actual *practices* coincide with (or are far removed from) citizens' expectations with regard not only to what a democracy is or how it should work, but mainly to how public policies are designed and implemented (by governments and their partners), and evaluated by citizens.

In what concerns citizenship, the correlation between "incomplete/flexible institutional design" and "quality of democracy" was considered from a twofold perspective.

On the one hand, not only citizens' rights *recognized* under democracy *protect* citizens, but also these rights *empower* them, in terms of offering them opportunities to struggle for new rights they may be currently lacking (Iazzetta 2004). On the other hand, and especially in contexts where there is democratic deficit, the institutional design as *unfinished business* (i.e., flexibility/incompleteness), instead of improving institutions' efficacy and enhancing social participation, has led to a situation where "rigidity" became the rule of the game, favoring a situation where rights are guaranteed only for the few. From this perspective, a consensus about the *quality of democracy approach* was, then, built over the interface *democratic regime - institutional performance - and citizenship.*

Such interface implied that the study of *political practices*[5], established on a daily basis, is a crucial approach to understand not only the flexible – or incomplete or rigid – aspects of public institutions, but, most importantly, to by what means citizens exercise their civil and political rights and obligations *with* their leaders, civil servants, decision-makers, and other agents, regarding the handling of public affairs (Vargas Cullell 2004).

In these terms, the idea that quality of democracy is not conceived as a general concept of the whole democratic system, but rather as the accumulated effect of *institutional performance,* and citizens' interactions on multiple fronts (Pérez-Liñan 1998) was, then, adopted.

Context-Dependent Variables: Institutional Design and Democratic Quality

Several scholars have argued that attempts of institutional design are at the heart of democratic politics.

Thus, in theories of institutional design, the notions of incompleteness and flexibility have been referred to as useful conceptual tools that help us deepen our knowledge about crucial aspects of political and policy processes in distinct institutional contexts.

Considering that these are all context-dependent concepts, there is always the possibility of drawing a distinction between different types of incompleteness/flexibility.

Intention

The motivation for creating a flexible/incomplete institution – The *intention* to create a flexible/incomplete institution in a democracy involves the analysis of the normative view regarding the purposes of the institution (policy statement, political manifestos, agitation by citizens or officials etc). Moreover, since the motivation to create an institution of that type (incomplete/flexible) may come from multiple sources – political parties, national government, citizens' associations, and/or diverse sectors working in a coalition – mobilizing political forces for the creation of flexible institutions is not an easy task. In fact, mobilization may be part of a political strategy aiming at the achievement of substantive goals related either to the delivery of public policies and/or to alter the distribution of political power in society. In any case, flexibility/incompleteness may offer the space to achieve particular goals that existing institutions perceive as being unable to accommodate for a variety of reasons – e.g. lack of capacity, lack of political commitment, control by elites who do not support popular movements etc.

Design

The analysis of formal rules that govern the institution – In relation to the *design*, institutional flexibility is positively associated with the extent of incompleteness (i.e., it is an inverse function of the level of completeness in the rules governing the institution). Since incompleteness is an *action*, it is perceived as an action of on-going design. On this matter, it is worth reminding not only that the very definition of *institution* (presented in this chapter) presupposed a permanent *on-going process* – which continues, and is broadly expected to continue, into an indefinite future[6]. Also, as contract theory tells us, since no one can specify for all possible eventualities, the on-going design is not only comprehensible, but also necessary.

Practices

Informal behaviours and conventions that emerge over time in the way actors within the institution act – In addition, according to the institutional theory, flexibility arises from the ways in which actors creatively respond to situations for which the formal rules are not adequate (Lowndes and Roberts 2013), and, in doing so, they promote *practices* that involve informal behaviors and conventions that emerge over time in the way actors within the institution act.

The quality of democracy approach, in turn, helps us clarify the effectiveness of specific policies in particular contexts, in terms of strengthening (or not) the quality of the political regime. The presumption here is that every democratization process entails a crucial element of institutional incompleteness – which involves, as we have seen, "expectations of indefinite endurance" (O'Donnell 2001). This statement, however, does not mean that institutional incompleteness prevails *only* in contexts of democratization. Instead, it is closer to affirm that "institutional completeness" is, in both cases (consolidated and nonconsolidated democracies), only an ideal. In other words, if the idea of incomplete institutional design is an "adequate" point of departure, *completeness* is hardly a point to be achieved anywhere. The point at stake is more directly related to the tension between the *ideal (as completeness)* and the *experience (as incompleteness)* – or, in Dahlian terms, between democratic *ideals and values* and the *reality* of liberal components of poliarchies[7] (Almeida 2004) – than to the very definition of completeness/incompleteness as "an act of power".

Also, the definition of quality of democracy has a normative component, which relates, in the first place, to the idea of democratic aspirations. For citizens to demand more quality from a democracy, their actions depend upon a set of capabilities (cognitive, organizational, moral) that are shaped by specific social contexts.

From this perspective, incompleteness, in terms of institutional flexibility, may be a "good thing", since it may help to increase citizens' capacities to interact in the *political life* established on a daily basis, in order to adjust the design of policies and the quality of public services to social needs.

Moreover, in political terms, *flaws* in the existing law, in the application of the law, in the relations between state agencies and ordinary citizens, in access to the judiciary are all important problems that indicate a "severe incompleteness of the state", especially in its legal dimension (Mendez, O'Donnell, and Pinheiro 1999).

This argument may well characterize ideas of incompleteness and/or incomplete institutional design in countries where democracy and its institutions are constantly being *reinvented* (as it seems to be the case of Brazil, at least, for the last 30 years).

In other words, in democratic transitional contexts, incomplete institutional design might bring about relevant *political* problems that affect the way institutions are (gradually) transformed – like the reinforcement of informality, which may therefore become the institutional "rule" (DaMatta 1986, 1987; O'Donnell, 2007).

Final Remarks

This chapter was based on an investigation developed in partnership between the Institute of Local Government Studies (INLOGOV – University of Birmingham), and the Centre of Public Policy Research (NUPPs – University of São Paulo). It described the building up of a conceptual framework to be applied in a comparative perspective, as well as the questions that emerged from our discussions about historical oriented notions – such as institutional rigidity, and incompleteness. It showed that, in political science, in general, and in public policy field, in particular, theoretical terms assume distinct functions in different contexts.

The analysis presented in this chapter in relation to incomplete/flexible institutions is twofold.

On the one hand, it demonstrated that in contexts where democracy is a tradition, and citizens' civil and social rights are largely widespread, incompleteness/flexibility take a different approach from that in which the democratic regime has been disrupted, and political institutions are constantly *being* reinforced, redesigned, and/or "completed". That is to say, whereas in the UK, institutional flexibility/incompleteness enhanced social participation and democratic procedures, and facilitated overall legal and political homogenization to policy production, in countries where the coexistence of more or less constitutional patterns with the subsistence of patrimonial kinds of rule prevail, flexibility/incompleteness may open space for decreasing credibility of political institutions as effectors of common good of their population. In this context, instead of improving social participation and policy efficacy, such notions may impose institutional rigidity and the insulation of decision-making – as scandals of its corrupt colonization have lately abounded.

On the other hand, it also showed that more *flexible* modes and channels of social participation, of political tolerance and of protest emerged as an attempt to respond to new (and old) demands expressed by the crisis of representation. In Brazil, that emergence took place after the democratization process of the 1980s.

These arrangements for governance, however – in spite of the fact that they are crucial components for both democratic regimes –, they bring about different impacts in old and new democracies.

In the UK, for instance, flexible and incomplete forms of organizational design have altered significantly public policy relationships, and these new complex arrangements have engaged citizens, politicians, civil society organizations, business representatives, the high bureaucracy of the state, and academic researchers in the debate, in order to shape and oversee the delivery of local policies operating at arm's length from elected officials (Skelcher 2012).

In Brazil, on the contrary, as well as in other Latin American countries, this new reality has produced and/or reinforced a contrast in cognitive levels about the understanding of what democracy really is. In this case, institutional flexibility and/or incompleteness tends to deprive social life of a sense of collective orientation, enhancing a phenomenon that O'Donnell named "low-intensity citizenship" (2005). Moreover, universalistic and equalizing policies that were designed to strengthen social participation are left floating in a fragmented and unequal society that, in turn, is loosely linked to a state that performs poorly its role as a rudder that would aim at giving some direction to collective orientation. By the same token, in contexts that lack democratic tradition, as the state at times seems to disintegrate in the banality of its incapacities of producing effective policies, politics itself tend to share that banality.

Therefore, mapping the terrain of argumentation that leads our discussion to an analytical strategy that aims to overcome current limitations in our fields (political science and public management), by identifying the conditions under which the dialogue between the states, their governments, and societies can be maintained (despite our cultural differences), is the challenge that we still have to face. As research teams, we aim, not to offer "new recipes" for the public administration communities, but rather useful possibilities that improve the credibility of our political institutions, the efficacy of our social policies, and the quality of our democracies.

Notes

1. This chapter is a result of a research ("New-Dem: Do flexible institutions enhance democracy? A comparative analysis of public governance innovations in Brazil and the UK") developed by the Centre of Public Policy Research of the University of São Paulo (NUPPs/USP) with the Institute of Local Government Studies (INLOGOV) of the University of Birmingham, from 2014 to 2016. It was financially supported by the São Paulo Research Foundation (FAPESP), under Grant numbers 2014/50221-6 and 2015/13026-3, and the University of Birmingham.
2. Marta Maria ASSUMPÇÃO-RODRIGUES, PhD in Political Science, University of Notre Dame, is Professor of Public Policy Management at the School of Arts, Sciences, and Humanities of the University of São Paulo, and senior researcher at the Centre for Public Policy Research of the same university. mmar@usp.br
3. José Veríssimo ROMÃO NETTO, PhD in Political Science, University of São Paulo, is a researcher at the Centre for Public Policy Research, and at the Centre of Metropolitan Studies of the same university, and at the Brazilian Centre for Analysis and Planning. joseverissimo11@gmail.com

4. By "path dependence", we refer to a theory in political science that focuses on how institutions come to constrain organizational life. It based on the historical-institutionalist approach, whose main assumption is that "history matters".
5. By "political practices", we refer to public policies that governments implement to provide social protection in several policy areas – such as security, sanitation, education, public health -, in order to expand citizens' rights (Mainwaring and Scully 2010).
6. As O'Donnell remind us, citizens, for instance, must get used to the fact that as elections "do not refer to a onetime event but to a series of elections that continue into the future" (O'Donnell 2004).
7. By liberal components of a polyarchy (Dahl 1971: 8) we mean the civil and political rights that would guarantee the exercise of a full-fledged citizenship.

References

Aguilar Villanueva, L. F. (2009) *Gobernanza y gestión pública*. México, DF: Fondo de Cultura Económica.

Almeida, M.H.T.de (2004) State, democracy, and social rights. In G. O'Donnell, J. Vargas Cullell and O.M. Iazzetta (eds.) *The quality of democracy: theory and applications*. Notre Dame, IN: University of Notre Dame Press.

Assumpção-Rodrigues, M. (2015) *Conceptualising quality of democracy*. Research Report n.2, NUPPs/USP, University of São Paulo.

Assumpção-Rodrigues, M. (2018) (org.) *Governança, qualidade da democracia e políticas públicas. Teoria e análise*. Rio de Janeiro, RJ: Editora UFRJ.

Avritzer, L. (2000) Democratization and changes in the pattern of association in Brazil, *Journal of Interamerican Studies and World Affairs*, 42(3), 59–76.

Avritzer, L., and Santos, B. de S. (2002) (orgs.) *Democratizar a democracia: os caminhos da democracia participativa*. Rio de Janeiro, RJ: Civilização Brasileira.

Avritzer, L. (2009) *Participatory institutions in democratic Brazil*. Washington DC: Woodrow Wilson Center Press.

Avritzer, L. (2017) *The two faces of institutional innovation: promises and limits of democratic participation in Latin America*. Massachusetts, MA: Edward Elgar Publishing.

Boniface, D. S. (2002) Is there a democratic norm in the Americas? An analysis of the Organization of American States, *Global Governance*, 8(3), 365–381.

Bovaird, T., and Loeffler, E. (2015) *Public management and governance*. New York, NY: Routledge.

Bresser Pereira, L. C., and Spink, P. K. (2006) (orgs.) *Reforma do estado e administração pública gerencial*. Rio de Janeiro, RJ: Editora FGV.

Brinks, D. M. (2010) Institutional design and judicial effectiveness: lessons from the prosecution of rights violations for democratic governance and rule of law. In S. Mainwaring and T. Scully (eds.) *Democratic governance in Latin America*. Stanford, CA: Stanford University Press.

Coelho, V. S. P. (2004) Considerações sobre o processo de escolha dos representantes da sociedade civil nos conselhos de saúde de São Paulo. In L. Avritzer, (org.) *A participação em São Paulo*. São Paulo, SP: Editora UNESP.

Dahl, R. (1971) *Polyarchy. Participation and opposition*. New Haven, CT: Yale University Press.

DaMatta, R. (1986) *O que faz o Brasil, Brasil?* Rio de Janeiro, RJ: Editora Rocco.

DaMatta, R. (1987) The quest of citizenship in a relational universe. In J. Wirth et.al. (eds.) *State and society in Brazil. Continuity and change*. Boulder: Westview Press.

DaMatta, R. (1995) Do you know who are you talking to?. In G. H. Summ (ed.) *Brazilian mosaic. Portraits of diverse people and culture*. Wilmington, DE: SR Books.

Diamond, L., and Morlino, L. (2005) (eds.) *Assessing the quality of democracy*. Baltimore, MD: The Johns Hopkins University Press.

Diamond, L., and Plattner, M. F. (eds.) (1997) *Consolidating the third wave democracies*. Baltimore, MD: The Johns Hopkins University Press.

Doimo, A. M. e Assumpção-Rodrigues, M. M. (2003) A formulação da nova política de saúde no Brasil em tempos de democratização entre uma conduta estatista e uma concepção societal de atuação política, *Política & Sociedade*, 3, 95–115.

Durose, C. et al. (2009) *Empowering communities to influence local decision making: a systematic review of the evidence*. London: CLG.

Evans, P. (1995) *Embedded autonomy. States and industrial transformation*. Princeton, NJ: Princeton University Press.

Feldman, M, and Khademian, A. M. (2007) The role of public manager in inclusion: creation communities of participation, *Governance. An International Journal of Policy, Administration, and Institutions*, 20(2), 305–324.

Filgueiras, F. (2018) Accountability, democracia e políticas públicas no Brasil. In M. M. Assumpção-Rodrigues (org.) *Governança, qualidade da democracia e políticas públicas. Teoria e análise*. Rio de Janeiro, RJ: Editora UFRJ.

Goodin, R. (1996) Institutions and their design. In R. Goodin (ed.) *The theory of institutional design*. Cambridge, UK: Cambridge University Press.

Gurza Lavalle, A. (2011) Após a participação. *Nota introdutória, Lua Nova*, 84, 13–23.

Hallerberg, M., Scartascini, C., and Stein, E. (2009) *Who decides the budget? A political economy analysis of the budget process in Latin America*. Cambridge, MA: Harvard University Press.

Huntington, S. (1991) *The third wave: democratization in the late 20th century*. Oklahoma, OK: University of Oklahoma Press.

Iazzetta, O. M. (2004) Introduction. In G. O'Donnell, J. Vargas Cullell and O.M. Iazzetta (eds.) *The quality of democracy: theory and applications*. Notre Dame, IN: University of Notre Dame Press.

Llanos, M., and Marsteintredt, L. (eds.) (2010) *Presidential breakdowns in Latin America: causes and outcomes of executive instability in developing democracies*. New York, NY: Palgrave Macmillan.

Levistky, S., and Way, L. A. (2010) *Competitive authoritarianism: hybrid regimes after the cold war*. Cambridge, NY: Cambridge University Press.

Linz, J., and Stepan, A. (1996) *Problems of democratic transition and consolidation*. Baltimore: MD: The Johns Hopkins University Press.

Lowndes, V., and Roberts, M. (2013) *Why institutions matter. The new institutionalism in political science*. London: Macmillan International.

Lowndes, V. (2014) *Conceptualising incomplete institutional design*. Research Report n.1. Birmingham: INLOGOV, University of Birmingham.

Mainwaring, S. (1999) *Rethinking party systems in the third wave of democratization: the case of Brazil*. Stanford, CA: Stanford University Press.

Mainwaring, S., and Hagopian, F. (2005) *The third wave of democratization in Latin America: advances and setbacks*. Cambridge: Cambridge University Press.
Mainwaring, S., and Torcal, M. (2006) Party system institutionalization and party system theory after the third wave of democratization. In R. S. Katz and W. Crotty (eds.) *Handbook of political parties*. London: SAGE Publications, 204–227.
Mainwaring, S., and Scully, T. R. (2010) Democratic governance in Latin America: Eleven lessons from recent experience. In S. Mainwaring and T.R. Scully (eds.) *Democratic governance in Latin America*. Stanford, CA: Stanford University Press.
Mainwaring, S., and Pérez-Liñán, A. (2013) *Democracies and dictatorships in Latin America*. Cambridge, NY: Cambridge University Press.
Melo, M. (1995) Ingovernabilidade: desagregando o argumento. In L. Valadares e M. Coelho (orgs.) *Governabilidade e pobreza no Brasil*. Rio de Janeiro, RJ: Civilização Brasileira.
Mendez, J., O'Donnell, G., and Pinheiro, P. S. (eds.) (1999) *(Un)Rule of law and the underpriviledged in Latin America*. Notre Dame, IN: University of Notre Dame Press.
Nunes, E. de O. (1984) *Bureaucratic insulation and clientelism in contemporary Brazil. Uneven state building and the taming of modernity*. Ph.D. Dissertation. Berkeley, CA: University of California.
O'Donnell, G., Schmitter, P., and Whitehead, L. (eds.) (1986a) *Transitions from authoritarian rule: prospects fo democracy. Southern Europe*. vol.I. Baltimore, MD/London: The Johns Hopkins University Press.
O'Donnell, G., Schmitter, P., and Whitehead, L. (eds.) (1986b) *Transitions from authoritarian rule: prospects for democracy. Latin America*. vol.II. Baltimore, MD/London: The Johns Hopkins University Press.
O'Donnell, G., Schmitter, P., and Whitehead, L. (eds.) (1986c) *Transitions from authoritarian rule: prospects for democracy. Tentative conclusions about uncertain democracies*. vol.IV. Baltimore, MD/London: The Johns Hopkins University Press.
O'Donnell, G. (1994) Delegative democracy?, *Journal of Democracy*, 5(1), 55–69.
O'Donnell, G. (2001) Democracy, law, and comparative politics. *Studies in Comparative International Development*, 36(1), 7–76.
O'Donnell, G. (2004) Human development, human rights, and democracy. In G. O'Donnell, J. Vargas Cullell and O.M. Iazzetta (eds.) *The quality of democracy: theory and applications*. Notre Dame, IN: University of Notre Dame Press.
O'Donnell, G. (2005) *Counterpoints. Selected essays on authoritarianism and democratization*. Notre Dame, IN: University of Notre Dame Press.
O'Donnell, G. (2007) Democratic theories after the third wave: a historical retrospective, *Taiwan Journal of Democracy*, 3(2), 1–10.
O'Donnell, G. (2010) *Democracy, agency, and the state: theory with comparative intent*. Oxford: Oxford University Press.
Olsen, J. P. (2013) The institutional basis of democratic accountability, *West European Politics*, 36(3), 447–473.
Plattner, M. F. (2005) A skeptical perspective. In L. Diamond and L. Morlino (eds.) *Assessing the quality of democracy*. Baltimore, MD: The Johns Hopkins University Press.
Pérez-Liñán, A. (1998) El estudio de la democracia en perspectiva comparada: nuevas preguntas, viejas respuestas, *Postdata*, 3, 221–241.

Pérez-Liñán, A. (2007) *Presidential impeachment and the new political instability in Latin America*. New York, NY: Cambridge University Press.

Pogrebinski, T. (2014) The impact of participatory democracy: evidence from Brazil's national public policy conference, *Comparative Politics*, 46(3), 313–332.

Przeworski, A. (1986) *Capitalism and social democracy*. New York, NY: Cambridge University Press.

Przeworski, A. (2010) *Democracy and the limits of self-government*. Cambridge, NY: Cambridge University Press.

Romão Netto, J. V. (2010) Estado, o pedagogo da liberdade: reformas das instituições político-administrativas do estado e cultura política no Brasil Império e República. *Tese de doutorado*. São Paulo, SP: Universidade de São Paulo.

Romão Netto, J. V. (2015) Gestão de políticas de cultura e qualidade da democracia. São Paulo, 10 anos de um modelo ainda em construção, *RAP. Revista Brasileira de Administração Pública*, 49, 1011–1038.

Skelcher, C. et al. (2005) The public governance of collaborative spaces: discourse, design and democracy, *Public Administration*, 83(3), 573–596.

Skelcher, C. (2012) What do we mean when we talk about "hybrids" and "hybridity" in public management and governance? *Working Paper*. Birmingham: Institute of Local Government Studies, University of Birmingham.

Skelcher, C., and Romão Netto, J. V. (2016) *Do flexible institutions enhance democracy? Theoretical aspects for a comparison of public governance innovations between Brazil and UK*. Research Report n.3. Birmingham/São Paulo: INLOGOV, University of Birmingham and NUPPs, University of São Paulo.

Stein, E., and Tommasi, M. (2005) Political institutions, policymaking processes, and policy outcomes. *A comparison of Latin American cases*. Washington DC: Inter-American Development Bank. Mimeographed.

Sullivan, H., and Skelcher, C. (2002) *Working across boundaries. Collaboration in public service*. London: Palgrave, Macmillan.

Vangen, S., and Huxham, C. (2003) Nurturing collaborative relations: building trust in inter organizational relations, *The Journal of Applied Behavioural Science*, 39, 5–31.

Vargas Cullell, J. V. (2004) Democracy and the quality of democracy: Empirical findings and methodological and theoretical issues drawn from the citizen audit of the quality of democracy in Costa Rica. In G. O'Donnell, J. Vargas Cullell and O.M. Iazzetta (eds.) *The quality of democracy: theory and applications*. Notre Dame, IN: University of Notre Dame Press.

Weber, M. (1978). *Economy and society*. 2 vols. Berkeley, CA: University of California Press.

4 The "glocal" elements in the linguistic atlas of Brazil

Suzana Cardoso & Jacyra Mota

This article presents aspects of the *Projeto Atlas Linguístico do Brasil* – ALiB (Linguistic Atlas of Brazil Project) with the purpose of describing its characteristics, highlighting its importance for the knowledge of the Portuguese language in its Brazilian modality, and highlighting cultural aspects related to the type of settlement, the people existing here before the arrival of the Portuguese, and those who were brought here as slaves or came as immigrants in search of new opportunities.

Linguistic Atlas of Brazil Project (ALiB)

The project for the elaboration of the ALiB, with regard to the Portuguese language, began in 1996 at the Federal University of Bahia (UFBA), at the *Seminário Caminhos e Perspectivas para a Geolinguística no Brasil* (Seminar on Pathways and Perspectives for Geolinguistics in Brazil). This seminar was attended by all the authors of Brazilian regional linguistic atlases, published or in progress, and by researchers associated with the area of geolinguistic studies in Brazil, as well as the dialectologist Michel Contini (Université Stendhal Grenoble)[1].

The main objective of the ALiB Project is to provide knowledge of the linguistic variation in Brazil, by presenting in linguistic maps the facts documented *in loco*, of 1.100 speakers distributed in 250 locations geographically located from north to south of the country. This knowledge will enable the dissemination of information concerning the characteristics that distinguish dialectal areas and sub-areas internally, and will show how Brazilian Portuguese is represented in relation to other Portuguese-speaking areas, especially that of Portugal.

It is also hoped that the dissemination of the ALiB data will permit greater visibility of the differences between the different areas. In this way, it will serve to match the teaching of Portuguese as a mother tongue, and as a second language, to regional realities and collaborate in the elaboration of dictionaries and pedagogical books. This will certainly contribute both to the reduction of the linguistic prejudices that affect, above all, the areas and speakers of lesser socioeconomic power

DOI: 10.4324/9781003225812-5

and to the wider dissemination of the reality of Brazilian Portuguese, particularly the regional lexicon and phonetic variation which distinguish it from the Portuguese of Portugal.

The ALiB Project is directed by a National Committee currently composed of 13 researchers from 10 Brazilian Universities.[2]

From a methodological point of view, the project adheres to the parameters of contemporary multidimensional Geolinguistics (cf. Thun, 2000. It classifies the informants systematically according to two age groups – the first, from 18 to 30 years, and the second, from 50 to 65 years – and of both sexes. In the state capitals, it also classifies them in two educational levels – the fundamental level, which comprises 9 years of schooling, and the university degree, with at least 16 years of schooling. Thus, the data present both diatopic (geographic) and diastratic (sociocultural) variation, as well as variation by gender.

For the documentation of the linguistic samples constituting the ALiB Project *corpus*, questionnaires specifically aimed at obtaining phonic data (including prosody), and lexico-semantic, morphosyntactic, and pragmatic data were used. Additionally, metalinguistic questions were included, and topics for recording semi-directed discourse and a reading text (Cf. Comitê Nacional Do Projeto ALiB, 2001).

The diversity in the type of questionnaire content permits the observation of register variation by facilitating comparison between speech registers more monitored by the speaker and more spontaneous registers.

The data are obtained through the application of indirect questions, which lead the interviewees to provide the most frequently used denominations in their locality, such as for the names that identify the rainbow: "Almost always, after a rain, a streak appears in the sky with colored stripes and curves (mimicry). What is the name of this streak?"[3] Some answers to this question are: *arco-íris, arco-celeste, arco-de-velho, arco-de-velha, olho de boi,* etc., produced with the characteristic phonetic variation of each area.

With respect to the results of the project, the first two volumes of the ALiB were published in 2014 (Cardoso et al. 2014a, 2014b). The first volume is introductory and provides information on the history and methodology of the ALiB Project. The second contains a total of 106 linguistic maps – phonetic, morphosyntactic, and lexical – with data from 25 capitals. The set of linguistic maps is preceded by information regarding the format of the maps, the history of the Brazilian capitals, the profile of the informants, the profile of the researchers who contributed to the constitution of the *corpus*, and ten introductory maps. Volumes 3–9 are in progress and a total of 20 volumes is projected. Volume 3 (with comments on the maps published in the second volume), Volumes 4 and 5 concern data from the capitals, whereas Volumes 6–9, present data from localities in the interior. In addition to the analyses for cartographic elaboration, the ALiB *corpus* facilitates the accomplishment of numerous partial analyses, which contemplate different levels of language study, from different points of view. To date, research on this *corpus* has yielded some 200 articles published in books or journals of the specialty, and 54 postgraduate theses, dissertations, and monographs. (Cf. Paim, 2016).

Also in progress at the Federal University of Bahia, the project for the elaboration of a *Dicionário de Dialetos Brasileiros* (DDB – Brazilian Dialect Dictionary), coordinated by Américo Venâncio Lopes Machado Filho, uses the ALiB *corpus*. The DDB Project emerged as a result of discussions between Brazilian and French researchers during the *Journée des Dictionnaires* (Dictionary Colloquium), in Salvador, in 2010, and was supported by computational resources developed at the *Lexiques, Dictionnaires, Informatique* – LDI (Lexicon and Dictionary Computation) Laboratory of the Paris 13 University (cf. Filho; Nascimento, 2016).

Other projects being developed by the ALiB team are the construction of a Computerized Database and a Sound Map, *on line*, coordinated by Daniela Claro Barreiro, and the *Cartography and Georeferencing in Geolinguistics Project*, coordinated by Ana Regina Ferreira Teles (cf. Ribeiro; Teles; Claro, 2016).

In the Brazilian context, the Linguistic Atlas of Brazil is an innovative project. At the international level, it stands alongside multidimensional atlases that differ from the unidimensional type typical of a long period of dialectal studies. It introduces methodological innovations that reflect advances in the formulation of the aims and purposes of Geolinguistics. In this sense, the Brazilian atlas adds social data of the informants to the diatopic variation traced on the linguistic maps. This permits the understanding of linguistic facts not explainable by the spatial diversity. In the treatment of data and its results, *The Linguistic Atlas of Brazil* resembles the *Atlas Lingüístico Diatópico y Diastrático del Uruguay* (ADDU – *Geographic and sociocultural linguistic atlas of Uruguay*) (Elizaincín; Thun, 2000), marking the beginning of a new perspective on dialectal studies in the American Continent.

The Importance of the ALiB for Geolinguistic Studies

Since its inception, the ALiB project has been characterized by its interconnection with related disciplines. And he is also able to articulate elements beyond the specific disciplines around him

It provides data that greatly surpass the limits of traditional smaller *corpora* and which permit transdisciplinary projection, touching in various aspects of the wider world. In the interdisciplinary field, the ALiB has contributed significantly to Brazilian Geolinguistics: it provides elements of renewal of methodology, stimulates research activity, and production in the field of Dialectology, fosters the celebration of academic congresses, and brings subsidies to other branches of language studies. According to Cardoso and Mota (2006), Brazilian Geolinguistics has served as a historical landmark, marking the beginning of a fourth phase of dialectal studies in Brazil. This fourth phase displays a significant increase in geolinguistic research production, with the appearance of linguistic atlases of regions and smaller domains. These studies contribute new perspectives, and thus allow the expansion of the dialogue between Geolinguistics and other fields of knowledge. Some such examples are the *Atlas Linguístico de Sergipe II* (Linguistic Atlas of Sergipe – II) (Cardoso, 2005), the *Atlas Linguístico do Paraná II* (*Linguistic Atlas of Paraná – II*) (ALTINO, 2007) and the *Atlas Fonético do Entorno da Baía da Guanabara* (*Phonetic Atlas of the Surrounding Area of Guanabara Bay*) (Lima, 2006).

In the *Atlas Linguístico de Sergipe II*, special attention was given to the presentation of data focusing geographic and gender factors. In this way, the relationship between linguistic data and social factors was introduced in a systematic manner, notably in the specific case of the sex of the speaker, showing how this link is constructed. The *Atlas Linguístico do Paraná II*, adopting a dialectometric perspective, revisits data that was not mapped in the *Atlas Linguístico do Paraná*) (Aguilera, 1994). The *Atlas Fonético do Entorno da Baía da Guanabara*), a micro atlas with phonetic-phonological data, covers four communities in the Rio de Janeiro Metropolitan Region, contemplating, in addition to spatial variation, both the gender and age of the informant.

Where data and information circulation are concerned, the ALiB is very much aligned with the computational and cartographic developments, using both dimensions as instruments to disseminate data collected in field research. Its linguistic maps are specified in several types, and the computational treatment of these same data in the form of a sound atlas. Examples of this type of technical presentation are the *Atlas Lingüístico Sonoro do Pará* (*Phonetic Atlas of Pará*), the first Brazilian linguistic atlas to incorporate acoustic information, based on ALiB methodology (Razky, 2004) and the *Projeto Atlas sonoro das línguas indígenas do Brasil* (*Phonetic Atlas Project of indigenous languages of Brazil*).[4]

For other branches of linguistic studies, the ALiB constitutes a representative database of information, concerning the lexicon – origin and nature of its composition, concerning the morphosyntactic configurations, and the representation of the phonetic paths followed by the Portuguese language in Brazil. These constitute important aspects for the knowledge and teaching of Portuguese as a mother tongue or as a second language. The variations in pronunciation, for example the realization of/R/and/S/in syllabic coda, which distinguish significant differences between regions (as noted in volume 2 of the ALiB), are phenomena of interest, particularly for the teaching of Portuguese as a second language.

The ALiB has also established interdisciplinary dialogues with other areas, such as the history of the settlement of the country, socioeconomic development of different regions or localities, and cultural aspects reflected in the language, as will be exemplified in section 2.

The ALiB in National and International Academic Environment

Interest in the results presented in the ALiB has led researchers to request data for the development of specific research work. Such a request by professors João Saramago (Lisbon, Portugal) and Xulio Sousa (Santiago de Compostela, Spain) has permitted them to carry out dialectometric analyses currently under development.

Exchanges with French universities have been developed by means of formal cooperation agreements between the Brazilian *Coordenação de Aperfeiçoamento de Pessoal de Nível Superior* (CAPES – Coordination for Improvement of Higher Education Personnel) and the Comité Français d'Évaluation de la Coopération Universitaire et Scientifique avec le Brésil (COFECUB – French Committee for Evaluation of University and Scientific Cooperation with Brazil), signed between Brazilian and French universities. The ALiB Project has participated in these cooperation agreements since 2009, together with the Paris 13 University, through Projects 651/2009 (already completed) and 838/2015 (ongoing). In 2014, French cooperation increased with the inclusion of the Paris Sorbonne 4 University), home to researcher Inès Sfar, participant in the CAPES-COFECUB Project, and host university of the *IV Congresso Internacional de Dialetologia e Sociolinguística* (IV CIDS – Fourth International Congress of Dialectology and Sociolinguistics).

Among the results of the CAPES-COFECUB agreements, the following are notable: a) the *Journée des Dictionnaires* colloquium, which led to the publication of *Os dicionários: fontes, métodos e novas tecnologias* ictionaries: sources, methods and new technologies) (Cardoso; Mejri; Mota, 2011) by Brazilian and French researchers; and b) the *Projeto de Estudo da Variação Lexical: teorias, recursos e aplicações* (VALEXTRA – Lexical Variation Study Project: Theories, Resources and Applications), under the coordination of Salah Mejri (Paris 13 University) and Marcela Paim (UFBA).

The VALEXTRA Project has as its main objective the survey of phraseology in Brazilian Portuguese, using the ALiB Project *corpus*, in order to provide data for the preparation of a bilingual dictionary. This will constitute an important work especially for the teaching of Portuguese as a foreign language, and for translation. It can also contribute to the study of lexical elements that reflect motivational and representative aspects of local cultures, as in the terms *chuva de pedra* "rain of stones" and *chuva de flor* "rain of flowers" to represent hail "precipitation of ice", *ponta de rama* "tip of the tree branch" for the youngest child, and *água que passarinho não bebe* "water that birds don't drink" for cane rum (*cachaça*), etc.

With regard to the *Asociación de Lingüística y Filología de América Latina* – ALFAL (Association of Linguistics and Philology of Latin America), in addition to the participation of individual researchers in the regular conferences, the ALiB is one of the 26 projects officially associated with ALFAL, since 2008. ALiB's participation in ALFAL is coordinated by two members of the National Committee. At the XVIII ALFAL congress in July 2017, at the National University of Colombia, 12 papers were presented based on the ALiB *corpus*.

Another type of exchange is established through the participation of researchers from foreign universities as co-supervisors of postgraduate academic projects, such as doctorates in co-tutelage and sandwich fellowships (*bolsa – sanduíche*).[5] In the first case, the doctoral candidate has two supervisors, one Brazilian and one foreign; in the second case, the candidate receives guidance from the foreign academic during the scholarship period spent at the foreign institution. The doctoral theses of Bassi (2016), through co-tutelage, and Rolo (2016), by sandwich fellowship, are notable examples of research comparing Brazilian Portuguese with that of other areas, based on ALiB data, or on linguistic samples documented using the same methodology.

In the doctorate by co-tutelage, Bassi (2016) compared data from the ALiB, recorded in Florianópolis and Rio de Janeiro, and data from Lisbon and São Jorge (Azores), documented with the same methodology of the ALiB. The aim was to verify the influence of the Azoreans in Brazilian areas where palatalized variants (chiantes) predominate in syllabic coda, in terms such as *estrada* "road", *casca* "bark, husk", *desvio* "deviation", and *mesmo* "same".

The results presented by Bassi (2016) confirm data previously analyzed by Furlan (1982), with respect to Santa Catarina, and ALiB data in which Florianópolis, Rio de Janeiro, Belém and Macapá stand out due to the greater frequency of palatalized variants. (Cf. Cardoso et al. 2014b, maps F05C1 to F05C4S).

Rolo (2016) researched two locations in Bahia (Bom Jesus da Lapa and Macaúbas) and two in Minas Gerais (Almenara and Itaobim) in Brazil, and São Jorge in the Azores, also using the same methodology as the ALiB. The analysis focused on the deletion of final unstressed vowels, in words like *dente* "tooth" and *estudo* "study". The analysis of the data verified the similarity of this phenomenon between these two areas, which can be explained by the presence of Azoreans during the settlement of Brazil. The settlement of the Brazilian localities studied was associated with the gold and diamond cycles, when many Portuguese, encouraged by the Crown, moved to these areas for the purpose of prospecting.

The Brazilian Portuguese in the ALiB *Corpus*

The Portuguese language, present in all the continents, is the official language of eight countries (Portugal, Brazil, Angola, Mozambique, Guinea Bissau, Cape Verde, São Tomé and Príncipe, and East Timor). It is the fifth most spoken language in the world, or the third, following English and Spanish, if the western countries are considered alone.

Brazilian Portuguese, like that of other regions beyond Portugal, has the characteristics of a transplanted language. This reflects not only the distance between Portugal and Brazil but also the coexistence of the Portuguese language in Brazil with other languages, indigenous or those of other peoples who came as slaves, or as migrants.

Thus, there are linguistic facts in Brazil that can be considered vestiges, conserving elements of an earlier phase of the language, no longer existent in Portugal, that also include modifications resulting from linguistic innovations occurring only in Brazil. The innovations were driven by the contact between Portuguese and other languages or resulting from an imperfect learning of the language of the settlers by speakers of other languages.

These facts, although already pointed out by other researchers through research in various regions and locations, are now systematically documented by the surveys of the 250 localities that compose the ALiB Project network.

Among the conservative features of Brazilian Portuguese, vestiges of earlier phases are found for example: a) at the phonetic level, the maintenance, in some Brazilian areas, of the dento-alveolar or sibilant pronunciation for the -s, in a syllable coda, and the timbre of the unstressed pre-tonic and final vowels; b) at the morphosyntactic level, the use of the possessive without the article; c) at the lexico-semantic level, the presence of the term *sarolha* "earth moistened by rain", registered in some areas of Brazil, as will be seen in the following sections.

Examples of innovations in Brazilian Portuguese are: a) at the phonetic level, the palatalized realization of the consonants/t, d/before [i] (as in *tia* "aunt", *dia* "day", *noite* "night", *tarde* "afternoon"); b) at the morpho-syntactic level, the generalized use of the verb *ter* "to have", in place of the verb *haver* "to be, exist", in phrases such as *aqui tem muitas árvores* "here there are many trees"; the gender variation for the adverb *menos* "less", as in *João tem menas força do que Manuel* "John has less strength than Manuel"; and instances such as *cinco degrais, cinco degrau* "five steps" alongside the standard form, *cinco degraus*; c) at the lexico-semantic level, the presence of words of non-Portuguese origin, from indigenous languages or from languages introduced later, by people of other origins present during the settlement of the country, in the first centuries.

Conservative Features

Phonetic Level

The following paragraphs highlight the dento-alveolar pronunciation of -s in the syllabic coda and the open timbre of the unstressed mid vowels.

The dento-alveolar or sibilant pronunciation for the -s, in a syllable coda (as in *casca* "bark, husk", *mesmo* "even", *mês* "month", *seis* "six"), in Brazilian Portuguese, differs from the palatalized variant heard in Portugal. For this reason, the Brazilian pronunciation was criticized in 1822, by the Portuguese grammarian Jerônimo Soares Barbosa, in the *Grammatica philosophica da língua portugueza ou princípios da grammatica geral applicados à nossa linguagem* (Philosophical grammar of the Portuguese language or principles of general grammar applied to our language). For this reason, the Brazilian pronunciation was criticized in 1822, by the Portuguese grammarian Jerônimo Soares Barbosa (1866 [1822]).

In Portugal, as Teyssier (1982, p. 55) observes, -s and -z in syllable final position "initially would have been sibilant, and in a later period, between the sixteenth century and the date of the first documented reference to the pronunciation (Verney, 1746), is when the *chiamento* 'palatalized pronunciation of the sibilant' would have been produced".[6]

The ALiB data document the conservation of the oldest sibilant pronunciation, mainly in twelve Brazilian capitals, five in the Northeast – São Luís, Fortaleza, João Pessoa, Maceió, and Teresina; three in the Southeast – Vitória, São Paulo, and Belo Horizonte; two in the Midwest – Campo Grande and Goiânia; and two in the South – Curitiba and Porto Alegre. In five other capitals, three in the North – Boa Vista, Rio Branco, and Porto Velho; and two in the Northeast – Aracaju and Natal – the palatalized pronunciation is restricted almost exclusively to the medial context of the word (*casca* "bark, husk", *mesmo* "even"), and the sibilant is maintained in the final position (*mês* "month", *seis* "six").

In only eight capitals, three in the North – Belém, Macapá, and Manaus; two in the Northeast – Recife and Salvador; one in the Southeast – Rio de Janeiro; one in the Center-West – Cuiabá; and one in the South – Florianópolis –, a palatalized sibilant equivalent to the current European Portuguese pronunciation is more often found. Among these, Rio de Janeiro, Florianópolis, Belém, and Macapá, mentioned above, stand out for registering the highest frequencies (cf. Cardoso et al. 2014b, linguistic maps F05 C1, F05 C2E, F05 C2G, F05 C2S, F05 C3, F05 C4E, F05 C4G e F05 C4S).

According to Silva Neto (1986, 1950), the palatalized pronunciation would have been brought to Rio de Janeiro by the Portuguese court, in 1808, from where it would have spread to most Brazilian areas, due to the prestige of the court and the city of Rio de Janeiro, the capital of the country at that time.

In relation to Florianópolis, the frequency of the palatal realization is generally attributed to the presence of Azorean immigrants, who came in the eighteenth century. Furlan (1995, 170) first made this observation, taking into account the extent of the feature in this area, and "above all, the almost total coincidence of the area of palatalization with that of the original Azorean settlement, which remained considerably isolated until 1970".[7]

According to Teyssier (1982, pp. 80–81), Brazilian Portuguese also displays conservative traits in relation to unstressed vowels, the latter being one of the most distinguishing features of European Portuguese. This author highlights: (a) the raising of the final mid vowel from [e] to [i], in in words such as *pass[i]* for *passe* "to pass", *noit[i]* for *noite* "night"; (b) the absence, in Brazilian Portuguese, of the mid central vowel [ə], characteristic of European Portugal, in examples such as *p[ə]cado* for *pecado* "sin"; (c) the conservation of the timbre [e, o] for the pre-tonic mid vowels, which are pronounced in Brazil as closed mid vowels in the Southeast, South, and Midwest.

Teyssier (1982, p. 81) notes as follows: "The Brazilian pronunciation concerning this item perpetuates once again the pronunciation of Portugal before the great phonetic mutations of the eighteenth century"[8].

With regard to the pre-tonic mid vowels, the ALiB contains eight maps that show the distribution of the pronunciation of closed [e, o] and open [ɛ, ɔ] in the capitals of the country, in words such as *terreno* "terrain, land" and *coração* "heart" (cf. Cardoso et al. 2014b, linguistic maps F01V1, F01V2, F01V3E, F01V3G, F01V3S, V01V4E, F01V4G, F01V4S).

The distribution of the open and closed pretonic mid vowels differentiates two major areas in the country – North and South – as Nascentes (1953 [1922] had already observed).

As for the word-final vowels, the occurrence of high vowels [i, u], as in *leite* "milk", *menino* "boy", is general throughout the country, as evidenced by ALiB data not yet mapped, with the exception of some areas such as Curitiba, where the pronunciations [e, o] are maintained.

Morpho-Syntactic Level

Examples of conservatism, at the morpho-syntactic level, such as "the possibility of using possessives without the article in cases where Portugal no longer does; e.g. *meu carro* 'my car'", also quoted by Teyssier (1982, p. 84), are extensively documented in the ALiB Project *corpus*, although they have not yet been systematically analyzed and, therefore, mapped.

Lexico-Semantic Level

In the lexical field, in 1963, the *Atlas Prévio dos Falares Baianos* (preliminary Atlas of Bahian Speech Varieties) registered the occurrence of *sarolha* for "land dampened by rain" in more than 50% of the network points. The term *sarolha* is an archaism, documented in the *Livro de Montaria* of D. João I (Pereira, 1918, p. 150) in the same sense, and no longer present in contemporary dictionaries of Portugal and Brazil. However, it is current in the Bahia area and alo in the State of Sergipe (cf. Ferreira et al. 1987; Cardoso, 2005). In Sergipe, *sarolha* is documented not only for moist soil, but also for two types of food – a type of *beiju* "thin pancake made with manioc flour" moistened with coconut milk, as listed in the *Dicionário Houaiss da Língua Portuguesa* of 2001, and a *farofa* "toasted and flavoured manioc flour" made with water and not with oil. The ALiB detects *sarolha* with its original meaning of "earth moistened by rain", in Northeastern states, attesting the presence of archaic traces in the lexicon of that region.

Innovations

Phonetic Level

The palatalized realization of the consonants/t, d/before [i] (as in *tia* "aunt", *dia* "day", *noite* "night", *tarde* "afternoon") is attested in volume II of the ALiB, in most Brazilian capitals. Seven capitals are distinguished only by the low occurrence of the palatalized variant: five in the Northeast – Natal, João Pessoa, Recife, Maceió, Aracaju – one in the South – Florianópolis – and one in the Center-West – Cuiabá (cf. Cardoso et al. 2014b, linguistic maps F06C1, F06C2E, F06C2G, F062S, F063E, F063G, F063S).

These palatalized variants were classified in the middle of the last century by Silva Neto as having of a stratified character, since, according to him, they would occur "in proportion to a lowering or increase of the social scale"[9] (Silva Neto, 1986 [1950], p. 162). However, in Brazilian Portuguese, currently, these palatalized pronunciations for/t, d/before [i] are prestigious, unlike the dento-alveolar, conservative pronunciation of Portugal, which is stigmatized by speakers of the palatalizing areas.

As for the lack of prestige or even the stigmatization of the non-palatal variants, the ALiB *corpus* provides interesting evidence. For example, there are instances in which the informant uses the palatalized variant in the answers to the questions of the phonetic-phonological questionnaire specifically directed toward obtaining this form, as in the words *tio* "uncle", *dia* "day". Yet the same informant begins to use the dental variant, later, when referring freely to some event.

Cases such as these are useful for verifying the value that speakers assign to the most frequent variant in their speech community. They bring to mind the well-known observation of Labov (1972), known as the "observer's paradox": the observer wants to document the more spontaneous and more frequent variant in the informant's speech, but to do so, uses questions that lead the informant to use the variants he/she considers more prestigious.

The greater prestige of some variants, to the detriment of others, is also generally explained by the influence of the media. Since 1956, following a decision taken at the *I Congresso Brasileiro de Língua falada no Teatro* (*First Brazilian Congress of Language spoken in the Theater*, 1958), the media in Brazil has adopted the most frequent variants in the areas of greater socio-cultural and economic power, and where the major television networks are located.

By making the variants of lesser prestige visible, the ALiB data can contribute to a greater acceptance of these variants, reducing the global importance of speech standardization and giving space to the still under explored glocal character of the language.

Morpho-Syntactic Level

Among the innovative morphosyntactic features, one can cite the generalized use in Brazilian Portuguese of the verb *ter* "to have", instead of *haver* "to be, exist", in cases such as "*Tem um rio grande que é maior e tem aquele rio que é derivado desse (...) que chama braço de rio*".[10] "There is a big river that is bigger and there is that other river that is derived from this (...) that is called the river's arm".

The high frequency with which the verb *ter* "to have" is documented with existential value, instead of the verb *haver* "to be, exist", in the Brazilian capitals, is presented in AliB, linguistic map M04 (cf. Cardoso et al. 2014b).

The innovations documented in Brazilian Portuguese may also be considered to include inflectional forms that deviate from the grammatical norm, such as *degrais*, as a plural form of the noun *degrau*, instead of *degraus* "steps" and *menas* occurring with feminine gender nouns in place of *menos* "less", as in *menas força* instead of *menos força* "less force".[11] *Degraus* and *degrais* are found in maps M01 (which shows the diatopic distribution in the capitals) and M01E (which contains diatopic-diastratic data). The forms *menos* and *menas* are listed in maps M03 (diatopic), M03E (diatopic-diastrática), and M03G (diatopic-diagenerational) (Cf. Cardoso et al. 2014b).

The forms *degrais* and *menas*, although also documented in the recordings of university-level informants, occur more frequently in the speech of individuals with lower levels of education (*escolaridade fundamental* "fundamental education"). The form *menas* is also observed in the recordings of the oldest informants (of the second age group), and registered in the notes accompanying the linguistic maps. The occurrence of the form *degrais*, recorded in 17% of university-level informants' speech, may be indicative of an ongoing change process that is evident in the ALiB mapped data.

Lexico-Semantic Level

The lexico-semantic level benefited from contributions by countless speakers of other languages and was enriched by the inclusion of denominations of Tupi and African origins, and items originating from languages of immigration or resulting from the contact between Portuguese and Spanish, in frontier areas.

The lexico-semantic questionnaire of the ALiB Project, with 202 questions, distributed in 14 semantic areas, sought to elicit a significant number of words in order to provide elements for a better understanding of linguistic variation in this semantic field.

The analysis of these words shows a large number of those identified in the dictionaries as Tupi origin or probable indigenous origin, such as: a) *macaxeira, mandioca, aipim* (cf. map L08), for a type of edible root; b) *carapanã* for a type of mosquito, present in the six capitals of the Northern region of the country – Rio Branco, Boa Vista, Manaus, Porto Velho, Macapá, and Belém (see map L14); (c) *tapuru* for a wrinkled, white beetle that attacks guava and coconut, was also documented in all the capitals of the North, and in four of the Northeast – São Luís, João Pessoa, Recife, and Maceió (see map L 13); d) *mangará* for the extremity of the flower head of the banana plant (see maps L07a and L07b), in the northern capitals, with the exception of Belém, and part of the Northeast – São Luís, Teresina, Fortaleza, Natal, João Pessoa and Recife; e) *morotó*, for the fly larva or a type of worm that appears in dung or rotten wood, registered in Salvador (see note to map L13).

The term *carapanã* has already appeared on the labels of products for combatting mosquitoes, as can be seen in the information on one of the types of insecticide: "*É eficiente para matar mosquitos (inclusive o mosquito da dengue), pernilongos, muriçocas, carapanãs, moscas e baratas*" "It is efficient for killing mosquitoes (including dengue mosquitoes), mosquitoes, muriçocas, carapanã, flies and cockroaches". All these denominations identify the same insect, the first two being in general use, and all are in map L14 of the ALiB.

Words of African origin include, for example, *galinha d'angola* and *guiné*, for a fowl similar to the chicken, of African origin. The first denomination was registered in all capitals, except for Maceió, and the second in six capitals of the Northeast – Natal, João Pessoa, Recife, Maceió, Aracaju, and Salvador (see map L11).

In addition to these examples, the documentation of *rapariga* (ALiB, v.2, maps L15Aa, b, c, e) for "prostitute", diverging from its original meaning as the feminine of *rapaz* "boy", which is maintained in the Portuguese of Portugal and of other areas. This has the effect that, in Brazilian Portuguese, reference to children of both sexes is expressed by the formula "*tenho dois filhos: um rapaz e uma moça*" "I have two children: a boy and a girl", whereby *moça* figures as the female counterpart of *rapaz* "boy". This avoids the linguistic form that, on the Brazilian side of the ocean, has acquired a new – and stigmatized – sense, that of "prostitute".

Immigration has played a significant role in shaping a collectivity, either through the transposition of ways of life, or by the introduction of names to identify what already exists in the area that receives the immigrant. An interesting example in the ALiB data emerges from the answers to question 177 of the Lexico-semantic Questionnaire that studies the names given to a fruit paste to spread on biscuits, toast, etc. In a certain area of southern Brazil, the response *chimia* (from the German *schmier*) occurred systematically – and not the traditionally used term *geleia*. The term *chimia* is, in fact, the word used by groups of German immigrants established in the region, who use the normal denomination in their dialect of origin.

Other denominations reflect the contact between Portuguese and Spanish, in areas close to the borders with the Hispanic countries (Paraguay, Bolivia, Argentina, and Uruguay). They include, for example, *pandorga* for a type of toy that flies in the wind, made of paper, "kite", documented in the capitals of the Midwest and South, while in other capitals the denominations are: *raia/arraia*, *papagaio*, *pipa*, *avião*, etc.

In the interrelationship between the different types of discourse — formal, relaxed, speculative, erudite – the use of the ALiB data is visible in citations in formal discourse, where *neblina* "mist, haze" and the phonetic variant *librina*, occur together with the lexico-semantic variants *neve*, *névoa, nevoeiro, neveiro, nuveiro*, all denominations for "fog, mist", documented in AliB, map L03 (Cf. Cardoso, 2017).

Sociocultural Aspects Revealed by the ALiB

Through the set of information that they gather, Linguistic atlases, and specifically, the *Atlas Linguístico do Brasil*, permit the development of research in different fields linked to cultural aspects that identify areas and ways of life in the Brazilian territory, as discussed in the following paragraphs.

In correlations between languages, particularly between the Romance languages, certain linguistic items documented in the ALiB suggest a distant connection. This connection is rooted either in the expansion of a denomination, or in an identification of a concept, perhaps genetic, in the creation of names to identify elements of the bio-social world. Thus, to name the rainbow there are denominations such as *arco-da-velha*, *arco-da-aliança*, *arco-da-velha-aliança*, *olho de boi*, that are found in other parts of the Romance language world. Why should *arco-da-velha* occur in Brazil and also in the Iberian Peninsula (Portugal and Galicia) as recorded in the *Atlas Linguarum Europae* (volume I, map 1.9), for example? Why should *olho de boi* occur both in Brasil and in France (also recorded on the aforementioned map 1.9)?

Such aspects that involve a correlation between languages are raised by this and other items documented in the ALiB.

The presence of a specific cultural influence of a country, of a region, on the customs and the *modus vivendi* of a collectivity, appears in the lexicon that linguistic atlases document. The atlases reveal uses that are no longer widespread, either because particular behaviors were replaced by others, or because cultural activities diminished and gave way to another type of presence. This is what has happened with the denominations for a certain type of coat that covers the neck and shoulders, protecting them from the cold, documented in areas of the interior of Brazil, that indicate a French presence, not only in the language, but in customs: *boá, cachecol, cachenê, fichu*. The AliB, volume 2, complements this aspect of the lexicon with two specific examples: *sutiã* and *ruge*.

The first example (Cf. Cardoso et al., 2014b, map L25) illustrates a type of cultural influence that has weathered time and maintained a continued use – the denomination *sutiã* for the female garment intended to support the breasts – notwithstanding the recording of other terms such as *corpete, porta-seio e califom*. Regarding the latter item, although Houaiss and Villar (2001) begin their lexical entry by declaring the etymology of *califom* to be dubious, they cite Castro (2001) who proposes a French origin: *califourchon*.

The second example from the ALiB (map L26), *ruge*, refers to a type of makeup, used to color the cheeks. The same map lists *ruge, blush* and *carmim*, the first two being prominent. These are two denominations with different origins: the first being French, and the second, English. The ALiB data, in the subsequent maps (maps L26E, 26G, and 26S), show that the item of French origin has been losing ground to the item of English origin. This is reflected in the generational data that show the prevalence of *blush* among younger age group of informants and a greater presence of *ruge* among the older age group. Where cultural influence is concerned, the ALiB data show very clearly how a denomination (e.g. *sutiã*) can persist despite the cessation of the dominant cultural influence of its area of origin, and reveal how another item *(ruge)* is overcome by the uses arising from new fields of political-cultural influence (*blush*).

Beliefs and religions display very clearly the reality of dealing with the supernatural, the unknown. In this regard, a first issue of interest concerns the transposition of terms used in religious rituals for use as common vocabulary. This occurs with the term *matinas*, referring to the first and also last hours of the day, which is documented in geolinguistic data from rural areas as representing *aurora*, representing daybreak and also dusk. In this case, the name with which, in the religious orders, the prayers made at exactly those times are identified it is necessary to bring to the common use and as a designative of a specific part of the day.

Other examples emphasize aspects that reveal markers of Brazilian cultural areas, such as the belief in the *boitatá* and the denominations for the devil, some of them of euphemistic character in order to attenuate the fear of those who pronounce them, and to remove the bad omens that they can bring. For example, the term *boitatá* was documented in Florianópolis, Santa Catarina, as "*uma bola de fogo que vinha rolando e daqui a pouco, se a pessoa não saísse, ele queimava a pessoa ...*" "a ball of fire that was rolling and in a short time, if the person did not leave, he burned the person ...". It is registered in dictionaries as an Indian myth, in the sense of "a snake of fire or light with two large eyes ..."[12], from Tupi etymology (Houaiss; Villar, 2001).

The abundance of denominations for "devil" is demonstrated in a thesis on this topic based on ALiB data from the Brazilian capitals (Costa, 2016). Although the most frequent denomination is *diabo*, the author lists 39 different names in response to question 147 of the ALiB's Lexico-semantic Questionnaire, which asks who is the being found in hell. Among these are metaphorical and metonymic forms such as *coisa ruim*, *inimigo*, *sujo*, *encardido*, *bicho feio*, *bicho ruim*, *chifrudo*, *enxofre*, *desgraça*, *troço*, used, generally to avoid the pronunciation of a dreaded name. There are also forms classified in dictionaries as brazilianisms, such as *capiroto*, and as azoreanisms, such as *cramulhano* (cf. Houaiss; Villar, 2001), and names such as *satangoso*, *sapirico*, *tibinga*, that are not in the more usual dictionaries.

Children's games reflect the diversity of motivations for each type of toy and the type of relationship established in each area. For example, the game played by drawing a shape in the form of numbered squares on the ground, and throwing a pebble and hopping. Volume 2 of the ALiB documents the following denominations: *academia, amarelinha, avião, cancão, caracol, macaca, macaquinho, macaco, maré, sapata* (Cf. maps L23 e L23a a L23e).

Another aspect worthy of mention is the relationship between the speaker of the language, the reality with which he/she is concerned and the choice of names to designate the elements of the speaker's world. Two cases illustrate the issue: designations for "entering menopause" and "conjunctivitis".

For "entering menopause", in the rural context, people create the following denominations by means of a metaphor constructed on the basis of their basic working tool (the machete): *amarrar o facão* "tie up the machete" and *quebrar o facão* "break the machete". There is no need for greater clarity to establish the comparison, because the absence of menstruation deprives the woman one of her greatest potentials, that of procreation.

For the second case, "conjunctivitis", the speaker seeks another type of solution: the description of the evil that he/she identifies as *dor d'olhos* "eye pain". Gradually, by differentiated phonetic processes and, above all, with the loss of linguistic motivation, different forms of naming have emerged from the same base: *dor de oio*, *dordolho*, *dordoio*, *dordoi.*

Concluding Remarks

The considerations presented here have shown that the ALiB Project is firmly rooted in the field of Brazilian linguistic studies, and specifically in the area of studies directed toward understanding linguistic variation. The ALiB has revealed new data by means of its particular approach to the description of the reality of Brazilian Portuguese. This contribution may be evidenced in the published materials of the project, in the diverse studies presented in at conferences, in specialist chapters of books and journal articles, and in the so far unexplored part of the project's vast corpus

Furthermore, the project has catalyzed the interest of specialists in the area who have been inspired by ALiB methodology to develop diverse studies, ranging from atlases of smaller domains, to specialized articles and chapters, and to the production of master and doctoral theses, as noted above.

The ALiB thus makes a very significant contribution to the consolidation of dialectal studies in Brazil, particularly by promoting the production of linguistic atlases of states or smaller domains.

In a broader perspective, in presenting the diversified Brazilian linguistic reality, the ALiB contributes to a greater acceptance of these regional variants. This fact demonstrates from the glocal point of view, one particular aspect of that reality still little explored. Knowledge of this reality tends to reduce the overall relevance of speech standardization.

The considerable interest in the ALiB *corpus*, shown by foreign researchers, points to the glocal character of the Project, evidenced in studies using ALiB data, conducted in different countries. Additionally, and stemming from this interest, are the foreign publications based on analyses of the materials collected by the Project and the requests for articles from the ALiB project for inclusion in international publications.

Notes

1. Michel Contini participates in three European linguistic atlas projects – the *Atlas Linguistique Roman* – ALiR (Linguistic Atlas of Romance languages), the *Atlas Linguarum Europae* – ALE (Atlas of European Languages) and the *Atlas Multimédia Prosodique de l' Espace Roman* - AMPER (Prosodic Multimedia Atlas of the Romance Language Regions).
2. Federal University of Bahia (Suzana Alice Cardoso, President, Jacyra Mota, Executive Director, and Silvana Ribeiro), Federal University of Pará (Marilúcia Oliveira), Federal University of Maranhão (Conceição de Maria Ramos), Federal University of Ceará and State University of Londrina (Vanderci Aguilera and Fabiane Altino), Federal University of Santa Catarina (Felício Margotti), University of Brasília (Abdelhak Razky) and Federal University of Lavras (Valter Romano).

3. Lexico-semantic questionnaire, question 17: *Quase sempre, depois de uma chuva, aparece no céu uma faixa com listras coloridas e curvas (mímica). Que nome dão a essa faixa?*
4. Project being carried out by Ana Suelly Arruda Cabral, Abdelhak Razky, Ariel do Couto e Silva, Jorge Lopes e Tabita da Silva, at the University of Brasília.
5. The Brazilian *bolsa sanduíche* fellowships provide doctoral students the opportunity to conduct their research both at the home university and at an overseas university.
6. Our translation of the original: "teriam sido inicialmente sibilantes, e, em época mais tardia, compreendida entre o século XVI e a data do primeiro testemunho (Verney, 1746), é que se teria produzido o chiamento"
7. Our translation. The original: "sobretudo a coincidência quase total da área da palatalização com a do primitivo assentamento açoriano, que permaneceu em acentuado isolamento até 1970".
8. Our translation. The original: "A pronúncia brasileira nesse ponto perpetua mais uma vez a pronúncia de Portugal antes das grandes mutações fonéticas do século XVIII".
9. Our translation. The original: "à proporção que se baixa ou se sobe na escala social"
10. Frase emitida por uma informante de nível universitário, de faixa etária II, em Salvador, a propósito da pergunta 01 do Questionário semântico-lexical do ALiB. The equivalente sentence using *haver* would be "Há um rio grande que é maior e há aquele rio que é derivado desse (...) que chama braço de rio".
11. Documentadas, respectivamente, como respostas às questões 16 e 32 do Questionário morfossintático (cf. Comitê Nacional Do Projeto ALiB, 2001).
12. Our translation. The original: "uma cobra de fogo ou de luz com dois grandes olhos ..."

References

Atlas Linguarum Europae (ALE). (1998). *Assen-Maastricht: Van Gorcum, 1983-1990*. v. 1–4. Roma: Istituto Poligrafico e Zecca dello Stato, v. 5.

Atlas Linguistique Roman (ALiR). (1996). Roma: Istituto Poligrafico e Zecca dello Stato; *Libreria dello Stato*, v. 1.

Altino, Fabiane Cristina. (2007). *Atlas Linguístico do Paraná II*. Londrina: Tese. (Doutorado em Estudos da Linguagem) – Universidade Estadual de Londrina.

Aguilera, Vanderci Andrade. (1994). *Atlas Linguístico do Paraná*. Curitiba: Imprensa Oficial.

Barbosa, Jeronymo Soares (1866 [1822]). *Grammatica philosophica da língua portugueza ou principios de grammatica geral*. 4. ed. Lisboa: Academia Real das Sciencias.

Bassi, Alessandra. (2016). *A realização da fricativa alveolar em coda silábica no português brasileiro e no português europeu: abordagem geolinguística*. 373 f. Lisboa: Tese (Doutorado em Linguística, em co-tutela) – Universidade Federal de Santa Catarina, Florianópolis; Universidade de Lisboa.

Cardoso, Suzana Alice Marcelino. (2005). *Atlas Lingüístico de Sergipe – II*. Salvador: EDUFBA.

Cardoso, Suzana Alice Marcelino. (2017). Discurso de Posse. *Revista da Academia de Letras da Bahia*. Salvador: Academia de Letras da Bahia, v. 1, n. 55, mar 2017, p. 255–277. In: https://academiadeletrasdabahia.files.wordpress.com/2017/03/revista-da-alb-55-para-o-site-20-02-2017.pdf. Acesso em 12.06.2018.

Cardoso, Suzana Alice Marcelino; MOTA, Jacyra Andrade. (2006). Para uma nova divisão dos estudos dialetais brasileiros. In: Cardoso, Suzana; Mota, Jacyra (Orgs.). *Documentos 2 – Projeto Atlas Linguístico do Brasil*. Salvador: Quarteto. p. 15–26.

Cardoso, Suzana Alice Marcelino; Mejri, Salah; Mota, Jacyra Andrade (Orgs.). (2011). *Os dicionários: fontes, métodos e novas tecnologias*. Salvador: Vento Leste.

Cardoso, Suzana Alice Marcelino *et al.* (2014a). *Atlas linguístico do Brasil*, v. 1. Introdução. Londrina: EDUEL.

Cardoso, Suzana Alice Marcelino *et al.* (2014b). *Atlas linguístico do Brasil*, v. 2. Cartas Linguísticas 1. Londrina: EDUEL.

Castro, Yeda Pessoa de. (2001). *Falares africanos na Bahia: um vocabulário afro-brasileiro*. Rio de Janeiro: Academia Brasileira de Letras; Topbooks.

Comitê Nacional Do Projeto ALiB. (2001). *Atlas lingüístico do Brasil*. Questionários 2001. Londrina: UEL.

Congresso Brasileiro De Língua Falada No Teatro, 1 (1958). Salvador. *Anais*... Rio de Janeiro: Ministério de Educação e Cultura.

Costa, Geisa Borges da. (2016). *Denominações para "diabo" nas capitais brasileiras: um estudo geossociolinguístico com base no Atlas Linguístico do Brasil*. 199 f. Salvador: Tese. (Doutorado em Língua e Cultura) - Universidade Federal da Bahia.

Elizaincín, Adolfo; Thun, Harald. (2000). *Atlas Lingüístico y diatópico del Uruguay*. t. I, fasc. A1. Kiel: Westensee.

Ferreira, Carlota et al. (1987). *Atlas Linguístico de Sergipe*. Salvador: UFBA-FUNDESC.

Furlan, Oswaldo. (1982). *Subsistência luso-açoriana na linguagem catarinense*. Rio de Janeiro, Tese (Doutorado) 420 f. Universidade Federal do Rio de Janeiro.

Furlan, Oswaldo (1995). Aspectos da influência açoriana no português do Brasil em Santa Catarina. In: Pereira, Cilene da Cunha; Pereira, Paulo Roberto Dias (Org.). *Miscelânea de estudos lingüísticos, filológicos e literários in memoriam Celso Cunha*. Rio de Janeiro: Nova Fronteira. p. 165–186.

Houaiss, Antonio; Villar, Mauro de Salles. (2001). *Dicionário Houaiss da língua portuguesa*. Rio de Janeiro: Objetiva.

Labov, William. *Sociolinguistics patterns*. (1972). Philadelphia: University of Pennsylvania Press.

Lima, Luciana Gomes de. (2006). Atlas Fonético do Entorno da Baía de Guanabara. 2006. Dissertação (Mestrado em Letras Vernáculas) – Universidade Federal do Rio de Janeiro, Rio de Janeiro.

Machado Filho, Américo Venâncio Lopes; Nascimento, Ivan Pedro Santos. (2016). *Sobre o projeto Dicionário Dialetal Brasileiro*. In: Cardoso, Suzana Alice et al. *Documentos 7*. Projeto Atlas Linguístico do Brasil – ALiB: 20 anos de história. Salvador: Quarteto. p. 175–190.

Nascentes, Antenor. (1953 [1922]). *O linguajar carioca*. Rio de Janeiro: Simões.

Paim, Marcela Moura Torres. (2016). *Projeto Atlas Linguístico do Brasil: a produção de 20 anos*. In: Cardoso, Suzana Alice et al. *Documentos 7*. Projeto Atlas Linguístico do Brasil – ALiB: 20 anos de história. Salvador: Quarteto. p. 191–253.

Pereira, Francisco Maria Esteves (Ed.). (1918). *Livro da montaria feito por D. João I, rei de Portugal*. Coimbra: Imprensa da Universidade. Disponível em: <http://www.archive.org/details/livrodamontariaf00johnuoft>. Acesso em: set. 2017.

Ribeiro, Silvana Soares Costa; Teles, Ana ReginaTorres Ferreira; Claro, Daniela Barreiro. (2016). Comissão de Informatização e Cartografia (CIC): dos primeiros passos às perspectivas atuais. In: Cardoso, Suzana Alice et al. Documentos 7. *Projeto Atlas Linguístico do Brasil – ALiB: 20 anos de história*. Salvador: Quarteto. p. 157–174.

Razky, Abdelhak. (2004). *Atlas Lingüístico Sonoro do Pará*. v. 1.1. Belém: CAPES/ UFPa/UTM.

Rolo, Maria do Carmo Sá Teles de Araujo. (2016). *Apagamento das vogais átonas finais* [i] *e* [u] *em áreas da Bahia e de Minas Gerais: aspectos históricos, geossociolinguísticos e acústicos*. 336 f. Tese (Doutorado em Língua e Cultura) – Universidade Federal da Bahia, Salvador.

Silva Neto, Serafim da. (1986 [1950]). *Introdução ao estudo da língua portuguesa no Brasil*. 5. ed. Rio de Janeiro: Presença,

Teyssier, Paul. (1982 [1976]). *História da língua portuguesa*. Tradução de Celso Cunha. Lisboa: Sá da Costa.

Thun, Harald. (1998). Atlas linguistique et variabilité – Introduction à la table ronde. In: XXIIe. CONGRÈS INTERNATIONAL DE LINGUISTIQUE ET DE PHILOLOGIE ROMANES, 22, Bruxelles, *Actes*...v. III, Tübingen: Max Niemeyer, 2000a. p. 407–409.

Thun, Harald. (2000). La géographie linguistique romane à la fin du XXe. siècle. In: CONGRÈS INTERNATIONAL DE LINGUISTIQUE ET DE PHILOLOGIE ROMANES, 22., 1998, Bruxelles. *Actes*... v. III, Tübingen: Max Niemeyer. p. 367–388.

5 The complexity and difficulties for crossing barriers to development of software used in international studies

Josiane Steluti & Dirce Maria Lobo Marchioni

Introduction

History of the Software

The software GloboDiet is a program which conducts a standardized interview and computer-based 24 h dietary recall (24-HDR), previously known as the EPIC-Soft software (Slimani et al, 1999). The program was developed for use in a large European multi-center study, namely the European Prospective Investigation into Cancer and Nutrition (EPIC). The study is one of the largest cohort studies worldwide, with more than half a million (521 000) participants recruited from 1992 to 1999 across 23 centers in 10 European countries (France, Italy, Spain, UK, Germany, the Netherlands, Greece, Sweden, Denmark and Norway) and followed ever since (Riboli et al, 2002). EPIC was designed to investigate the relationships between diet, nutritional status, lifestyle, and environmental factors, and the incidence of major cancer sites, cardiovascular disease, type 2 diabetes, mortality, and healthy ageing (Bingham & Riboli, 2004).

The GloboDiet has a structure and interview interface implemented to optimize the standardization of the dietary interview, i.e 24-hour diet recall. The structure of the interview in the software includes four main steps (Figure 5.1): 1) general information; 2) quick list of consumed food items; 3) description and quantification of foods and recipes; and 4) description and quantification of dietary supplements (Slimani et al, 1999).

Initially, the program was used as a reference calibration method among the centers participating in the EPIC study. The rationale for this approach was to use 24-hour dietary recall as a reference method to adjust the country-specific food questionnaire of improving the detection of the real relationship between diet and cancer or other chronic diseases (Slimani et al, 2002a). Thus, the EPIC-Soft was adapted for each

DOI: 10.4324/9781003225812-6

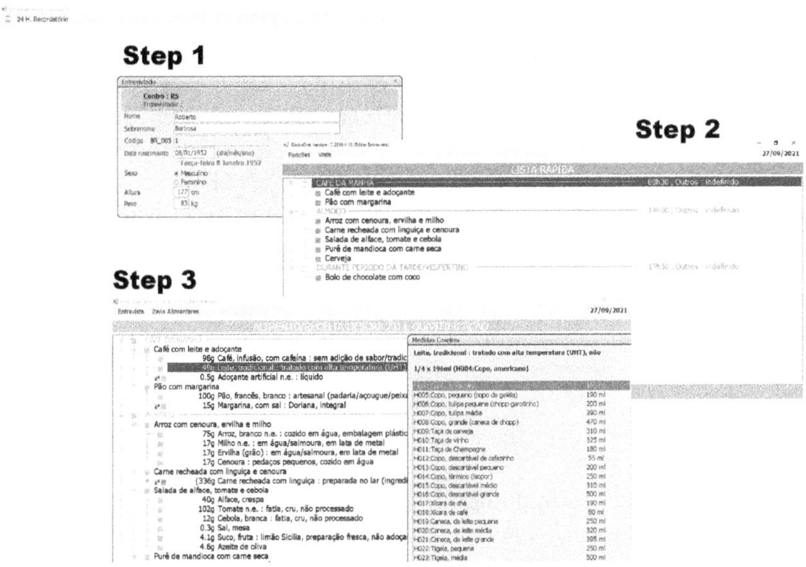

Figure 5.1 The structure of the interview in the Brazilian version of GloboDiet software

participating country and translated into nine languages that allowed to standardize dietary interviews between the EPIC centers (Slimani et al, 1999).

Expansion Initiative for Other Studies

There was growing interest in the harmonization of monitoring and surveillance systems to assess among other things the impact of health programs at the European Union level. The status of diet was one of measures evaluated. Initially, the guidelines were established using existing sources such as food balance sheets, household budget surveys and national monitoring surveys. Although all provide useful information, such data do not provide reliable estimates at the individual level either. Thus, the dietary intake data from the various national dietary intake surveys in Europe had limited comparability and could not be pooled.

In this context, the project "European Food Consumption Survey Method" (EFCOSUM) was settled within the framework of the EU Programme on Health Monitoring, in special in the context of pan-European dietary monitoring surveys (Slimani et al, 2002b). The main purpose of this project was to evaluate the comparability of existing national food intake data and propose a new study design for conducting food intake surveys in the countries of the European Union in a comparable manner (Brussaard et al, 2002). The EFCOSUM established that repeated

24-hour diet recall (24-HDR) measurements should be conducted as a dietary assessment method across countries. The EFCOSUM group then considered the possibility of using the existing computerized 24-hour dietary recall software, i.e. GloboDiet software (Ocké et al, 2011).

However, a series of updates were needed to optimize the usability of the GloboDiet software in the European countries. Thus, the European Food Consumption Validation (EFCOVAL) Consortium was created to continue the work initiated by EFCOSUM. This project aimed to further develop and validate the use of repeated 24-HDRs using GloboDiet software for the dietary assessment and potentially hazardous chemicals for surveillance purposes relevant to health and safety policies in Europe (Slimani et al, 2011). Afterwards, many studies concluded that GloboDiet software appeared to be adequate to describe the intake of protein, potassium, fish, fruit and vegetable in European populations (Crispim et al, 2011).

The GloboDiet methodology has been applied to different countries and study contexts for the collection of standardized dietary data for epidemiological, surveillance, and monitoring purposes. Currently, 19 European countries have already been using the software (Park et al, 2015). In addition, pilot initiatives have been carried out in other regions worldwide such as Latin America, Asia, and Africa. Korean, Mexican, and Brazilian versions of GloboDiet have been completed, and road maps for their local feasibility, validation and implementation are advanced (Park et al, 2015; Bel-Serrat et al, 2017). In Africa, an inventory, conducted as a prerequisite of any implementation, highlighted a lack of comparable dietary assessment methods and support infrastructure for research across the 18 countries represented, and elucidated specific needs and obstacles for implementation (Aglago et al, 2017).

The Proposal of the Brazilian Version

The successful opportunity of adapting GloboDiet to different cultural contexts in Europe provided its extension to Latin American countries. Nowadays, in Latin American countries, national studies on food and nutrition are using different methods of dietary intake assessment. The need to obtain dietary intake data is evident altogether with the possibility of international comparability in order to provide common global and regional databases for research and support of public health policies (Bel-Serrat et al, 2017). Thus, the International Agency for Research on Cancer (IARC) proposed the Latin America DIETary Assessment Project (LA-DIETA) that would develop a method of obtaining dietary intake on highly standardized proceedings, by using GloboDiet and its structure in Latin American countries.

Therefore, Brazil was one of the countries selected to start the LA-DIETA project in collaboration with the IARC because the country constituted a center of excellence and accumulated experience in national dietary surveys, and it provides a complex scenario in terms of population and food patterns.

Based on the success of the European experience, the use of the GloboDiet methodology to other regions, such as Latin America, was considered important to allow a global comparison and contrast of dietary intake. This possibility entailed an excellent opportunity to articulate the research and dietary monitoring conducted in Europe with other regions and populations. Besides, it would provide overall interpretation of dietary exposures and the outcomes related to them and further comparative and contrasting analyses. Furthermore, Brazil, more particularly, could greatly benefit from this software, since the country still showed some gaps in dietary information that could easily be collected and monitored allowing comparability at the national and international levels.

Software Development Phases in Different National Extensions

The development for the GloboDiet version for each country starts with the update or creation of 70 interrelated files on consumed foods, recipes, quantification methods, coefficients for calculation of consumed amounts and supplement list. Thus, some files are common to all countries participating in this international project, and other files are partly country specific files. In the first step, the common databases are generated, prepared, and translated into the respective native language. In a second step, country-specific GloboDiet databases have to be updated or created. It is necessary to describe and quantify foods, therefore, recipes and supplements are prepared by the local dietitians and nutritionists. This includes updating or creating list of foods and recipes, household measures (HHM), thickness of different types of bread, shapes, brand names, synonyms, pictures (food, recipe, HHM, portions), probing questions specific to foods and others (Bel-Serrat et al, 2017; Steluti et al, 2019).

In Brazil, around the same 70 databases, comprising common and country-specific databases, were adapted and translated into Brazilian Portuguese (Bel-Serrat et al, 2017; Steluti et al, 2019). The IARC GloboDiet database library contains all the common reference databases which apart from facilitating standardization among versions also serve as a starting point for the development and customization of new GloboDiet versions (Slimani et al, 1999). The description uses a facet-descriptor approach which allows the standardization of the level of details in order to describe foods and recipes in a comparable manner within and between countries. In addition, country-specific files were developed for each country, to acknowledge and integrate the country-specific features and pre-existing available data considering the food patterns, socio-cultural, ethic, and geographical origins. First, food and recipes list were defined based on European food lists, and additional food and recipes were included using frequencies of intake in national and/or regional dietary surveys. Second, all the identified foods and recipes were listed in the country-specific language and in English and classified according to the adapted GloboDiet classification (food and recipe (sub) groups).

The Complexity and Peculiarities regarding the Software Adaptation in Brazil

Brazilian food and recipe lists were built based on European food lists and using local data that reported frequencies of intake in national and/or regional dietary surveys in Brazil, specifically the 2008–2009 Brazilian National Dietary Survey (IBGE, 2011) and the "Inquérito de Saúde em São Paulo" (ISA) (Fisberg & Marchioni, 2012; ISA, 2017). All the identified foods and recipes were listed in the country-specific language, i.e. Brazilian Portuguese, and in British English and classified according to the adapted GloboDiet groups and subgroups. The same situation occurred in other countries which did not have British English as first language. So, the files related to the software had two versions, one of them in Brazilian Portuguese and the other in British English. The final Brazilian list included 1754 food items and 355 recipes. For vegetables, fruits and fish, scientific names were included to simplify their identification and appropriate classification (e.g. "açaí", a fruit grown in Brazil – its scientific name is *Euterpe oleracea*). Another example, in Brazil, due to the high reported intake and large number of recipes that incorporate "cassava" (*manihot esculenta*, a root native to South America), new subgroups to classification, i.e. "Cassava", were created under the recipe groups "Based on vegetables". Previously, the software only had subgroups that led to classify the recipes "based on vegetables" in the subgroups legumes, potatoes, other vegetables and mixed salads. Therefore, recipes, such as those with "tapioca" and "farofa" (both recipes are prepared with cassava flour) could be misclassified.

Another important issue was about the innumerous names for the same food considering the different regions of the country. To give an example, the Brazilian pepper ("pimenta rosa") has 17 shadow names (SH) identified (Figure 5.2). It is important to mention that shadow names also represented similar names with small spelling differences and particularities due to misspelling. So, the researchers elaborated a shadow list that included more than 300 food items of which 11% had five or more shadow names to facilitate the search and avoid mistakes during the dietary interview (Bel-Serrat et al, 2017).

The questions to be asked on different characteristics of foods and recipes, named "facets", were selected from the IARC GloboDiet database

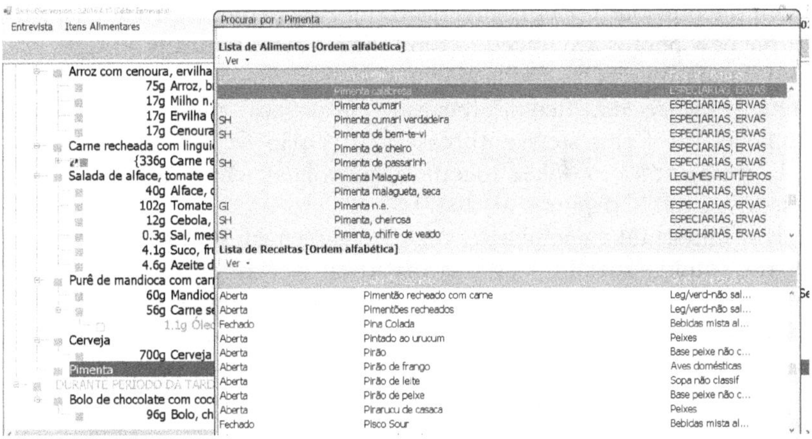

Figure 5.2 Examples of shadow names (SH) to the Brazilian Peppers available in the search system of the software

library. Sixteen GloboDiet facets for foods (e.g. source, physical state, cooking method, preservation method, sugar content, flavored/added component, fat content, salt content, etc.), and three for recipes (e.g. recipe known, recipe production, and recipe brand name) were selected for describing Brazilian foods and recipes. The predefined available answers, called "descriptors", in the database library for each facet were evaluated by the Brazilian teams and those not applicable to the country were discarded, more particularly, the descriptors of flavored components. In addition, missing descriptors necessary to describe the food items were identified and added into the common files and into the IARC library. Total of 57 new descriptors were found to be lacking in the GloboDiet library and coded for the Brazilian version. The majority of these new descriptors were listed under the facet "Flavored/added component". Fifty-three descriptors were identified, most of them referring to local tropical fruits such as "guarana", "açaí", "buriti", "cupuaçu", and others (Bel-Serrat et al, 2017). These flavors are commonly mentioned for ice-creams, juices and other desserts in Brazil.

The same modifications and adaptations were observed in the quantification methods of foods and recipes. These adaptations in the quantification method were required in order to estimate with accuracy the portion sizes consumed by the population. The quantification method included household measures (e.g. glass, cup, spoons etc.), photos, shapes (i.e. surface area equivalents), and standard units (e.g. can, bottle, one fruit etc.). The Brazilian picture book contains sets of photos of foods and recipes, pictures for spreads (e.g. the amount of butter spreads in the bread), Brazilian household measures, and shape of real size (e.g. slice of bread or pie). Only 24 new photos were created and 69 were taken from the previous GloboDiet versions for other countries. The majority of the new photos are linked to foods and recipes typically from Brazil, and standard units for foods item (e.g. portion sizes of "feijoada" (bean stew) – Figure 5.3, "feijão" (beans), "carne seca" (dried meat), "vaca atolada" (Beef ribs stew with cassava), "baião de dois" (fresh black-eyed peas and rice cooked together), chocolates (chocolates), biscoitos (cookies), etc) (Crispim et al, 2017).

It is important to emphasize that the agreement on the final content of the country-specific files was discussed weekly during joint online meetings organized between IARC team and Brazilian researchers. The IARC team included researchers, research assistants, project assistants

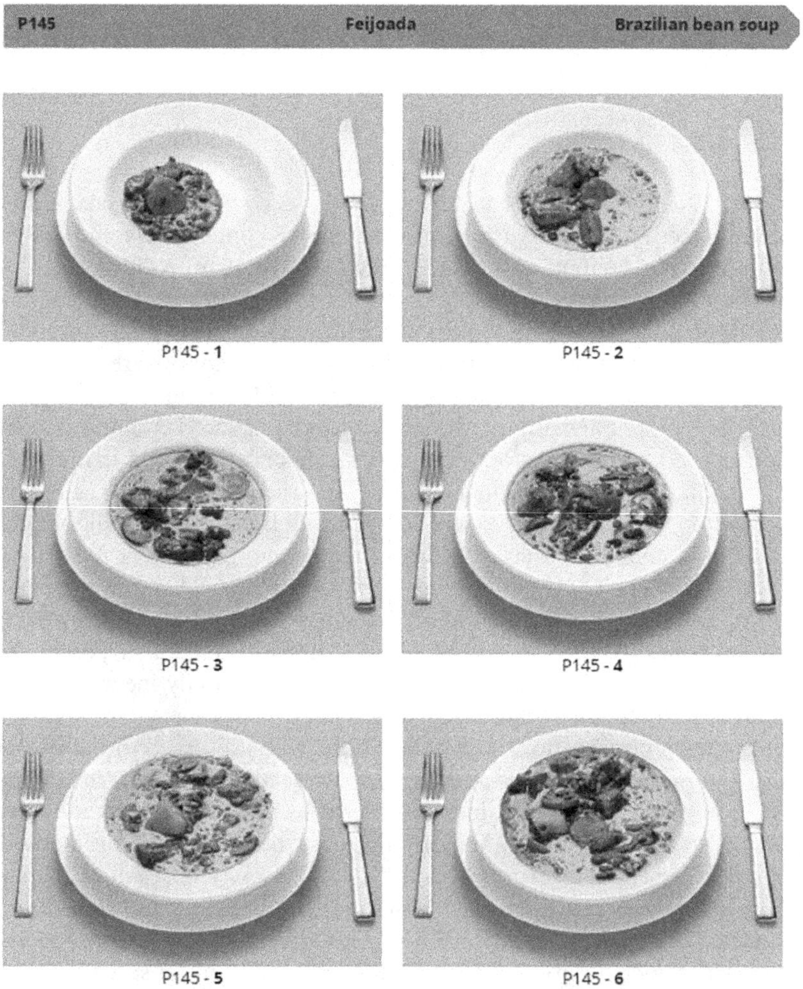

Figure 5.3 Portion sizes of the recipes typically from Brazil ("Feijoada")

and technical assistants, while the Brazilian team involved researchers, PhD students, nutritionists, and nutrition undergraduate students.

The Main Limitations

The development of the Brazilian version of the software was not an easy activity although many original files could easily be adopted. The proceedings included: 1) translation of the common files (e.g. food and recipe classifications, facets and descriptors for foods and recipes, probing questions etc.) and 2) customization of country-specific files (e.g. food and recipe lists, synonym lists, brand name lists etc.). The discussions were frequent mainly to explain the reason for the modifications on files which had already been standardized across countries. Brazilian researchers needed to justify these alterations based on national and local research studies on nutrition, such as Brazilian's cookbooks, photos produced by research group at Federal University of Paraná and search sites available in the country. However, these modifications have always been accepted and incorporated into the GloboDiet database library.

One of the main alterations that brought resistance by researchers from IARC was the inclusion of many recipes in the software database. The IARC team was reluctant because, according to them, the recipes only needed some changes in ingredients. They suggested that interviewers should not create so many new recipes in the software; they just needed to replace the ingredients. Nevertheless, the Brazilian researchers reported that the country had a huge diversity with regard to cooking. The recipes contain different ingredients, cooking methods, specific names, and portion sizes; therefore, changing one ingredient would not enable the identification or recognition of that recipe by the Brazilian population.

Thus, it is clear that software development does not only require to be directly translated into the Brazilian Portuguese language nor just following standardized guidelines. It demands an adaptation and customization of the files with respect to the different linguistic, socio-cultural, ethnic and geographical roots and, more particularly, to represent, in detail, food patterns in this country.

The Finalization and Use of the Software Brazilian Version

The Brazilian version of the software is still in the initial stages of testing for feasibility and validation by using biomarkers. So, the access to software is still restricted because they await the results of these studies. However, it is expected that free access to the software will soon be provided to all researchers who will want to carry out further studies that assess the dietary intake of the Brazilian population.

In order to facilitate access to researchers who are interested in the software, an user manual was elaborated for the Brazilian version of the GloboDiet software, which includes all the characteristics of the software as well as recommendations on how to conduct interviews. The manual has been translated from the original material provided by IARC, and adapted according to the information regarding the country-specific databases. The final user manual developed covered the following topics: I - Introduction, II - The interview and computer-based 24-hour dietary recall in the GloboDiet, III - Recommendations before beginning of the interview, and IV - Procedures for entering an interview. In addition, the Brazilian researchers adapted and translated Appendix 1 - Glossary of facets and descriptors (Steluti et al, 2019). This glossary contains all the definitions and applications regarding the facets and descriptors that defined in software Brazilian version. The facets (e.g. descriptors) were: source (e.g. beef, cow, goat, pork etc.); physical state (e.g. liquid, reconstituted from powder, powdered, flesh etc.); cooking method (e.g. roasted/baked without fat, barbecued without fat, boiled in water, steamed etc.); preservation method (e.g. canned, deep frozen, salted, smoked etc.); sugar content (e.g. non sweetened/no sugar added, sweetened with sugar, sweetened with artificial sweetener etc.); flavored/added component (e.g. chocolate, coffee, caramel, vanilla, nuts etc.); fat content (e.g. whole, half skimmed, skimmed/no fat etc.); and sodium content (e.g. normally salted, salt/sodium reduced, sodium restricted etc.).

Finally, a tutorial was recorded as a demonstrative video in which the information contained in the user manual was summarized. The video shows the interview conducted in the software Brazilian version. For the elaboration of this demonstrative video, a pre-elaborated script was designed, and it included part of 24-hour dietary recall interview, i.e. a structured interview intended to capture detailed information about all foods and beverages consumed by the individual in the past 24 hours (NIH/NCI, 2018). A specific program was used to record the software screen and audio at the same moment and, as a consequence, to generate a file in digital extension. In addition, an eLearning course was developed to train and certify researchers and health professionals for using of the Brazilian version of GloboDiet software.

Conclusion

The software GloboDiet is a program which conducts a standardized and computer-based dietary interview. Initially, the software was developed and adapted for European countries, after; different versions were carried out in other regions worldwide including Brazil. Despite the fact that there is a common structure and interview interface implemented by countries to optimize the standardization of the diet recall interview in the GloboDiet software, the researchers needed to face the complexity and particularities in order to develop the software Brazilian version respecting the different linguistic, socio-cultural, ethnic, and geographical features, in special, the national food patterns.

The processes of software development in different national extension included the translation and update of common files to all countries, in addition, the creation of country specific files. However, difficulties were experienced. All files related to food and recipes were listed in two languages (Brazilian Portuguese and British English). New foods (such as vegetables, fruits and fish) were included. Besides, new subgroup was created due to the high reported intake and large number of recipes from cassava, i.e. a root native to South America. Other relevant issue was the several shadow names added to the software to facilitate the identification of food in whole country respecting the regions differences. Lastly, other modifications and adaptations were conducted such as deletions/inclusions of food description and customization of quantification methods (e.g. household measures, portion sizes photos of foods and recipes etc.).

All these steps were discussed during joint online meetings organized between IARC team and Brazilian researchers, and were needed to development a software version that will be used to provide comparable food intake data within and between countries which are essential for monitoring and surveillance systems to assess risk and study the diet – disease associations.

References

Aglago EK, Landais E, Nicolas G, et al (2017). Evaluation of the international standardized 24-h dietary recall methodology (GloboDiet) for potential application in research and surveillance within African settings. *Glob Health*; 13(1):35.

Bel-Serrat S, Knaze V, Nicolas G, et al (2017). Adapting the standardised computer- and interview-based 24 h dietary recall method (GloboDiet) for dietary monitoring in Latin America. *Public Health Nutr*; 20(16):2847–2858.

Bingham S & Riboli E (2004). Diet and cancer–The European prospective investigation into cancer and nutrition. *Nat Rev Cancer*; 4:206–215.

Brussaard JH, Kearney J & Johansson L (2002). Rationale and methods of the EFCOSUM project. *Eur J Clin Nutr*; 56(Suppl 2):S4–S7.

Crispim S, Fisberg RM, Almeida CCB, et al (2017). Manual fotográfico de quantificação alimentar – Curitiba: Universidade Federal do Paraná. 2017. 147 (http://gupea.ufpr.br/?page_id=176) [Accessed on: Jun 28, 2020]

Crispim S, Geelen A, Souverein OW, et al (2011). Biomarker-based evaluation of two 24-h recalls for comparing usual fish, fruit and vegetableintakes across European centers in the EFCOVAL Study. *Eur J Clin Nutr*; 65(Suppl 1):S38–S47.

National Institutes of Health (NIH)/National Cancer Institute (NCI). Dietary Assessment Primer, Instrument Profiles. National Institutes of Health, National Cancer Institute. (https://dietassessmentprimer.cancer.gov/) [Accessed on: Jan 15, 2018]

Fisberg RM, Marchioni DML (2012). Manual de Avaliação do Consumo Alimentar em Estudos Populacionais: a Experiência do Inquérito de Saúde em São Paulo (ISA). São Paulo: Faculdade de Saúde Pública, Universidade de São Paulo; 2012.

Inquérito de Saúde de São Paulo (ISA) (2017). Avaliação do Consumo Alimentar. http://www.fsp.usp.br/isa-sp/ (Accessed on: Nov 2017).

Instituto Brasileiro de Geografia e Estatística (IBGE) (2011). Pesquisa de Orçamentos Familiares 2008–2009. Análise do Consumo Alimentar Pessoal no Brasil. Rio de Janeiro: IBGE.

Ocké MC, Slimani N, Brants H, et al (2011). Potential and requirements for a standardized pan-European food consumption survey using the EPIC-Soft software. *Eur J Clin Nutr*; 65(Suppl 1):S48–S57.

Park MK, Park JY, Nicolas G, et al (2015). Adapting a standardised international 24 h dietary recall methodology (GloboDiet software) for research and dietary surveillance in Korea. *Br J Nutr*; 113(11):1810–1818.

Riboli E, Hunt KJ, Slimani N, et al (2002). European Prospective Investigation into Cancer and Nutrition (EPIC): study populations and data collection. *Public Health Nutr*; 5:1113–1124.

Slimani N, Casagrande C, Nicolas G, et al (2011). The standardized computerized 24-h dietary recall method EPIC-Soft adapted for pan-European dietary monitoring. *Eur J Clin Nutr*; 65(Suppl 1):S5–15.

Slimani N, Kaaks R, Ferrari P, et al (2002a). European Prospective Investigation into Cancer and Nutrition (EPIC) calibration study: rationale, design and population characteristics. *Public Health Nutr*; 5:1125–1145.

Slimani N, Valsta L; EFCOSUM Group (2002b). Perspectives of using the EPIC-SOFT programme in the context of pan-European nutritional monitoring surveys: methodological and practical implications. *Eur J Clin Nutr*; 56(Suppl 2):S63–S74.

Slimani N, Deharveng G, Charrondière RU, et al (1999). Structure of the standardized computerized 24-h diet recall interview used as reference method in the 22 centers participating in the EPIC project. European prospective investigation into cancer and nutrition. *Comput Methods Programs Biomed*; 58(3):251–266.

Steluti J, Crispim SP, Araujo MC, et al (2019). Tecnologia em Saúde: versão brasileira do software GloboDiet para avaliação do consumo alimentar em estudos epidemiológicos. *Rev Bras Epidemiol*; 23:E200013.

6 Spaces of encounter and misencounter between researchers and local people in interdisciplinary and transdisciplinary studies in fishing villages

Charbel N. El-Hani & Rosiléia O. de Almeida

Introduction

Fishing communities hold important cultural assets for conserving natural resources in riverine, estuarine, and marine ecosystems. Around the globe, predatory fishing techniques have been threatening the survival of life in the oceans, wetlands, estuaries, and other water ecosystems and, consequently, have affected many ecosystem services. However, several fishing communities still maintain traditional knowledge and practices that can lead to sustainable use of natural resources[1], but are themselves threatened throughout tropical regions.

A team of Brazilian researchers, in collaboration with Colombian, Dutch, North-American, Italian, Namibian, and Singaporean researchers, have been carrying out interdisciplinary and transdisciplinary studies in two fishing villages in the North shore of the state of Bahia, Brazil, in the estuary of a large river (Itapicuru), in the municipality of Conde: Siribinha (ca. 500 inhabitants) and Poças (ca. 700 inhabitants). From these studies, we are developing a model for research and innovation in and with traditional communities, involving a transnational team of researchers and multi-stakeholder relations with the communities and local authorities. The goal of the project is to build, through this model, a knowledge base integrating scientific, practitioner, and traditional knowledge that can be used to promote intercultural education and community-based management in the fishing villages, while also fostering capacity-building among researchers, teachers, fishermen/women, and local authorities for managing the processes of construction and use of such knowledge base.

In the North shore of Bahia, fishing communities are gradually disappearing due to the growth of the tourism industry and declining catches resulting from the impact of overfishing, pollution, and other environmental threats. In many estuaries of the region, mangroves have been destroyed by the growth of human occupation or severely threatened

DOI: 10.4324/9781003225812-7

by changes in the riverine and estuarine systems, environmental contamination, and other impacts. However, in the Itapicuru River estuary, despite impacts, mangroves are still significantly conserved, as shown by the abundant presence of sensitive species such as the rufous crab hawk (*Buteogallus aequinoctialis*, locally known as gacici, near threatened, IUCN Red List).[2]

The local communities in the estuary use the mangroves for more than a century at least, based on their wealth of fishing knowledge and practices, which emerged historically as a cultural product from combined native South American and Portuguese influences (Ott, 1944). In Siribinha, for instance, at least 10 different fishing techniques are used. Each technique is a repository of knowledge made concrete in fishing practice. Despite historical changes suffered by these bodies of knowledge and practices in the last three decades, sometimes making them less sustainable, this is still a living fishing culture, with young fishermen and fisherwomen being recruited and knowledge flowing from grandfather/mother to son/daughter to grandchildren, and a number of sustainable techniques still preserved.

In our studies, a transdisciplinary and transnational team gathering 13 senior and 15 junior researchers puts to use 25 years of experience working with fishing communities, ethnobiological research, educational studies, and conservation efforts to support the conservation of both fishing culture and estuarine ecosystems.[3] The team also includes local stakeholders, 9 basic education teachers, 5 fishermen, and 1 environmental technician.

We are building a university-community partnership aiming at five goals: (i) to carry out studies on the fishing communities' bodies of knowledge and practices, especially ethnobiological and ethnoecological knowledge; (ii) to engage in collaborative work with local teachers to do research, innovation, and development of educational proposals for intercultural education as dialogue between school knowledge and traditional knowledge; (iii) to mediate collective processes of community members' reflection on their cultural heritage, as a way of pondering about the relations between their past, present, and future, and musealizing elements of their territory in order to make their cultural legacies visible for visitors who go there attracted by beach tourism (eventually leading to the development of an ecomuseum in their territory); (iv) to develop a participatory process for empowering the community for participating in environmental conservation plans being developed by local authorities in the Itapicuru River estuarine ecosystems, so that they can be heard in decision-making and engage in community-based conservation that respect their ways of living and generate Integrated Conservation and Development Projects (ICDP) (Gavin et al., 2015); and (iv) to derive, from the field studies, a general model for inter- and transdisciplinary research and innovation in traditional communities, which builds bridges between the natural sciences and humanities, scientific research and social development, and academic and other forms of knowledge.

We have been framing this research and innovation model in terms of a set of epistemological, methodological, and ethical criteria. Here are some examples of these criteria:

i One of the goals of the model is to build an approach to intercultural education in schools located in traditional communities based on the idea of a dialogue between different ways of knowing and forms of knowledge (owing to Paulo Freire's education for emancipation; e.g., Freire, 1970/1987).[4] As an epistemological background, we approach the nature and relations of different ways of knowing and forms of knowledge (say, traditional, scientific, technical, practical) from a non-relativist, but pluralist perspective (see, e.g., El-Hani & Mortimer, 2007; El-Hani, Silva-Filho & Mortimer, 2014). This means each one is recognized in its own validity, and with its own epistemic criteria for assuming what is valid and non-valid knowledge. At the same time, it means ways of knowing and forms of knowledge are demarcated to the extent that is possible, as diverse ways of solving different kinds of problems (without hierarchizing them in any absolute term).

ii All educational innovations are ideally constructed within a community of practice (CoP) (Lave & Wenger, 1991; Wenger 1998) engaging researchers and teachers as peers, even though it takes time and the right conditions for a CoP to develop from collaborative work. As conditions for CoP development, teachers' knowledge and practices, wishes, and aspirations are all the time recognized and valued in collaborative work, and vertical hierarchical relations are constantly under a critical lens. In concrete terms, this means an ethics of research in which teachers get truly involved in research, having a voice in proposing research goals and tools, and being involved in data gathering and analysis, and are always authors of research articles and other products alongside with the researchers. It also means an ethics for innovation in the classroom, in which we strive for maintaining the teachers' central place in the educational process, empowering them as innovators in their classrooms. Ideally, the final innovation is an emergent transdisciplinary product in which we can see how elements coming from both teachers and researchers intermingle into a cohesive teaching approach.

iii Research should bring return to the community while taking place, not afterward, and this return should be indeed valuable, i.e., clear benefits should ensue, while care is exercised to avoid bringing harm. That's why the approach necessarily combines research with innovation in education, social museology, participatory conservation processes. Other ways of bringing return to the community follow from the next principle.

iv We should be open to include, as research and innovation goals, desires of the community that may not be – at least not at first – our own goals and desires. For instance, we came to the communities to develop ethnobiological/ethnoecological studies and educational research and innovation, but ended up striving to pursue goals related to museology and conservation as an outcome of this principle.

Border Crossing across Disciplines, Knowledge Systems, Languages, and Cultures

This model directly links our work with plurilingual, intercultural, inter/transdisciplinary, and translational research. On the one hand, it engages us in intercultural dialogue with the members of the communities. Despite the fact that this is not a plurilingual relationship, as all the involved stakeholders speak Portuguese, we should not neglect that we need to bridge over differences that show some convergence with the challenges faced by speakers of different languages when they have to truly engage with each other in a joint meaning-making relationship. We should consider at this juncture a key aspect of our work, namely that we have been engaged for a long time in efforts to both understand the causes and attempt to bridge the so-called research-practice or knowing-doing gap (El-Hani & Greca, 2011, 2013; Pardini et al., 2013; Bertuol-Garcia et al., 2018, 2020), that is, the distance between relevant work produced by academic researchers and decision-making and action in social circumstances where that knowledge is useful, sometimes even crucial.[5] In the research and innovation model developed we also tackle two additional gaps, those between natural sciences and humanities, and between scientific and other forms of knowledge, like teachers' practitioner knowledge, and traditional knowledge. To bridge over these gaps, we need to build processes that engage the stakeholders put into relation in the complex quests of inter-ontological, inter-epistemological, and inter-axiological dialogue. That is, the stakeholders involved in these inter- and transdisciplinary processes are delving into circumstances that show similarities with speakers of different languages, dealing with their differences in worldviews, values, and knowledge conceptions in order to understand each other and engage in mutual meaning-making processes. To show why it is needed to deal with these complex intercultural dialogues, consider that, in a democratic vision, discourses, and practices affecting a whole socioecological system should be the outcomes of social negotiation and collective learning, and that in any socioecological system there are distinct ontologies, epistemologies, value systems among the several actors composing a network of social relationships. Thus, social negotiation and collective learning, as a democratic necessity, will always engage the stakeholders in a given socioecological system in inter-ontological, inter-epistemological, and inter-axiological dialogue.

Surely, we are not experts in language studies and cannot tackle these complex dialogues across gaps from a perspective informed by these studies. Rather, we deal with these issues from the standpoint of other bodies of literature, such as works on intercultural communication. Consider, for instance, Broome's (1991) proposal of meaning-productive relational empathy in intercultural communication. From this perspective, intercultural communication demands the development of shared meanings through social interaction among stakeholders, involving "a series of successive approximations to the other's point of view during social interaction" (p. 241). New meanings are produced through these interactions, supporting relational empathy and mutual understanding across cultural barriers (and across ontological, epistemological, axiological gaps). This means that such interactions are productive of novel, intercultural understanding. The outcome can be regarded as transdisciplinary, not only interdisciplinary, since it aims to be not only an integrated body of academic knowledge but to also engage with knowledge built by other stakeholders, such as teachers and fishermen/women.

In our work in and with traditional communities, we attempt, in sum, to engage in meaningful relationships that build mutual trust and understanding, producing new meanings from the several approximations to one another's points of view through collaborative and participant work, which can support forms of intercultural dialogue. Even though we are all Portuguese-speaking people in these sustained relations, we see similarities with plurilingual relations that are worth considering when we deal with glocal perspectives.

A second aspect of our work brings us into actual plurilingual perspective, as the model of research and innovation we are developing in the work in and with fishing communities is currently proposed to be applied in the Mrõtidjam village, in the Xikrin Indigenous Land Trincheira-Bacajá, in the municipalities of Altamira, state of Pará, Brazil, and in Ju|'hoansi and! Kung communities from Tsumkwe, in Namibia. This is still in the beginning but challenges posed by the different languages spoken in the project are already under our critical lens. The easiest ones are those posed by the fact that the project engages speakers from several different languages which can only count on English as a shared language. There are certainly problems in appealing to English to communicate among us, but for our transnational team of glocademics this is the most feasible strategy now. In the meantime, we have efforts for using Portuguese for communication, especially in fieldwork, as many of the stakeholders do not speak English. For some team members who are not Portuguese-speaking, this means language learning, even though part of the foreign researchers already speaks this language. But, as we move into working in Xikrin, Ju|'hoansi and! Kung communities, the challenge will become harder for almost everyone in the team, since these communities speak languages that most of the researchers do not master.[6] As fieldwork is still to begin in these communities, we are yet to see the prospects and limitations of carrying out our several lines of investigation there, but we already know we will need to rely on the expertise of those in the research team that can speak the languages, at least at first, and this will bring additional difficulties to the inquiry process.

We are still beginning to analyze the data we gathered in ethnographic, educational, and conservation studies, and, thus, we are not in a position yet to report and discuss these results here. Our intention is to discuss from our field notes and through self-reflection by three team members some spaces of encounter and misencounter[7] between the community members and ourselves. This is a first step in an ongoing reflection that will subsequently engage the other members of the research team and stakeholders from the local communities. In the next section, we will briefly elaborate on spaces of encounter and misencounter between researchers and local people.

Spaces of Encounter and Misencounter between Researchers and Local People

Anne Toomey (2016) discusses encounters and misencounters between environmental scientists and local people in Madidi National Park, in the Bolivian Amazonia. She builds on theories of contact, which examine encounters and relations between different groups and cultures in diverse settings, in order to explore questions that are surely relevant to our own work: What kinds of spaces exist (or emerge) when scientists carry out research? What types of encounters and misencounters occur between people in these spaces? What are the implications for scientists looking to bridge the knowing-doing gap?

Pratt (1992, p. 7) introduced the expression "contact zone" to refer to "social spaces where disparate cultures meet, clash, and grapple with each other, often in highly asymmetrical relations". For her purposes, the contact zone was "the space of imperial encounters", in which "peoples geographically and historically separated come into contact with each other and establish ongoing relations, usually involving conditions of coercion, radical inequality, and intractable conflict" (p. 8). However, the concept of contact zone seems useful to describe and explain several other social spaces of encounter and misencounter, where subjects and groups previously living separately are co-present, at a point where their trajectories intersect, as it happens when research teams work within and/or with traditional communities. One key advantage of this perspective is that it emphasizes how subjects are constituted in and by their relations to each other in the space where they coexist, interpreting these relations in terms of interactions, interlocking understanding and practices, and typically asymmetrical power relations, in order to conceive how reciprocal constitution takes place.

Encounters in the contact zone entail a double- or even multi-edge relationship, as shown by Pratt in her analysis of how travel and exploration writing produced "the rest of the world" for European readerships, while at the same time producing Europe's evolving conceptions of itself in relation to what was not European. While travel writing encoded and legitimated the aspirations of economic expansion and imperial powers, Pratt shows us that it also undermined those aspirations at certain points. Although the imperial metropolis tends to imagine itself as determining the periphery, *e.g.*, through civilizing missions or powers of development, it also tends to blind itself to the fact that the colonies also have the power to determine the metropolis itself. At the same time, the so-called peripheries have the power to transform and refract how they are represented under the eyes of the colonizers, tinkering with the colonizers' codifications of their reality. Their authors claim, revise, reject and transcend those representations and the ways they depict them and the places they inhabit (Pratt, 1992, p. 4).

Other scholars have described contact zones as spaces of "heterogeneous and unequal encounters" in which different peoples, values, worldviews, and knowledge can rub up against one another, leading to "new arrangements of culture and power" (Tsing, 2005, p. 5). These are spaces of "friction", to use Tsing's metaphor: "A wheel turns because of its encounter with the surface of the road; spinning in the air it goes nowhere. Rubbing two sticks together produces heat and light; one stick alone is just a stick" (p. 5). The metaphor of friction reminds us that in encounters across difference it is not only unequal relations of power leading to dominance that take place. New arrangements of culture and power emerge from transcultural connections, even when part of the actors comes with their imperial eyes (Pratt, 1992). The effects of encounters across difference are not only compromising but can also be empowering: we should be attentive not only to the conflicts, but also to the "possibilities of friction" (Tsing, 2005, p. 18).

These spaces of contingent contacts and encounters are surely never neutral. Rather, they are deeply laden with power, but as they unfold they change power itself. Thus, what happens within them cannot always (perhaps not even often) be anticipated and/or controlled (e.g., Tsing, 2005; Haraway, 2008; Toomey, 2016). For this reason, while such encounters may be "fraught with contestation and conflict", they also contain the potential for "connection, empathy and contract" (Sundberg, 2006, p. 239).

These concepts, with their accompanying images and metaphors, contribute to deepen thinking about the different spaces that are generated when researchers bridge the knowing-doing (or research-implementation) gap and engage with social practice. When researchers come to the field as we came, certainly not looking for any colonial encounter, and even entitled to question our power positions and relationships with local people, it is still worth thinking of the space of contact with local communities from those theoretical grounds. We are looking for connection and empathy, not for contestation and conflict, but, as relations between researchers and local people are certainly laden with power and are unequal, contestation and conflict can and often will happen, and even neocolonial attitudes may emerge (Rist & Dahdouh-Guebas, 2006). To be aware of unequal power relations is important not only to prepare ourselves better for conflicting relations, but to put ourselves under a critical lens that comes not only from outsiders; rather, as we recognize power and inequality in our relations, we should put ourselves under our own critical lens. That's why self-reflection becomes all so important in such processes of contact and encounter.

Moreover, the space between research and practice, between knowing and doing does not get simply dissolved once we intend to bridge upon it, but is rather resignified, becoming a space of encounter where new things may be possible. Encounters across difference are never one-sided, but multi-edge. While we work together with the community members, we produce a representation of them for ourselves, and they do the same about us, and in this landscape of mirrors these representations are all too important because they help shaping the relationships themselves. But it is more than that: as we build the relationships and the accompanying representations, we change our conceptions of ourselves, and the community members may also undergo such changes. By meeting in the contact zone, we ourselves become others. It is this identity transformation through meaningful relationships that a self-reflective perusal, a sort of ethnography of ourselves, may – no doubt imperfectly – bring to the surface. A self-reflective effort can both help and be potentialized by engaging in intercultural encounter and communication that may lead to meaning-productive relational empathy (Broome, 1991).

But, before going on with the argument, couldn't we be hampering all possibilities of such mutual engagement and communication by reinforcing unequal power relations through simply calling the communities "traditional". After all, as Sousa Santos and Meneses (2009) cogently argue, the designation "traditional" played a specific role in European colonization. This colonization process entailed an epistemicide that strived for systematically destroying any social practice of knowledge production that seemed to be against the colonizers' own interests. The outcome was a severe decrease in cultural, political and epistemological diversity in the world, a real waste of social and cognitive experiences, many of which could be precious now. When social epistemic practices survived the epistemicide, they were subjected to a dominant epistemological norm that was originated in Western Europe but was projected as universal and absolute. The reference to those practices as "traditional" (as opposed to European "modernity", "enlightenment"), or "local", or "contextual" comes in the wake of this colonizing attitude.

Thus, it becomes a key issue for any transdisciplinary process engaging academic researchers and traditional communities, especially if there is hope of knowledge integration, to be aware of and cautious about the possibility of (re)producing (neo)colonial projects. For this purpose, to recognize both prospects and limits of transdisciplinary integration is an important asset, as well as supporting the ontological and epistemological self-determination (Viveiros de Castro, 2009) of the indigenous communities when knowledge integration is not possible (Ludwig, 2016). The same can be said of the importance of being cognizant and critical of diverse attitudes one may take in relation to other cultures and knowledge systems, some of which are neocolonial and detrimental to indigenous people (Rist & Dahdouh-Guebas, 2006). Inter-ontological, inter-epistemological, and inter-axiological dialogue can be framed then as projects of resistance to epistemicide and subjugation, if accompanied by a fierce criticism of any attempt to deny the validity or self-determination of ways of knowing and forms of knowledge other than the Western ones.

We do not think, however, that one will be necessarily engaged in neglect or subjugation by calling these ways of knowing and forms of knowledge "traditional", or "local", or "contextual". Truly, these designations played a subjugation role in colonization processes, but nowadays they seem to gain new tones and meanings that may give them roles, indeed, in resistance processes. Notice, for instance, that in the literature on ecology and conservation "contextual" and "local" are taken as qualifications that show the undeniable value of the knowledge of traditional communities (e.g., Pierotti & Wildcat, 2000; Ludwig, 2016). Or that "traditional" became a designation that qualifies, through a series of laws, communities to access to rights and empowerment, to the extent that one of the researchers in our team (Viviane de Souza Martins, pers. communication) reported the resistance in a meeting of a community leader to an argument deconstructing the idea of "traditional community", precisely due to the perception that to be traditional has become to be entitled to forms of empowerment and resistance. Surely, this deserves more exploration, but for the moment being we keep using the terms "traditional", "local", "contextual" from a critical standpoint and with an empowering intention.

When meeting in the contact zone, both researchers and communities get entangled in emergent processes that can bring new perspectives to representation, contestation, conflict, empathy, connection, and collaboration. This demands continuous effort to build multi-edge relational processes that engage people and groups in interpersonal relationships that effectively bridge the worlds of research and community practice (Pain et al., 2011). As Pratt (1992, p. 7) reminds us, contact zones open up not only possibilities but also perils, requiring an adventurous attitude, which we should have when we face the difficult task of thinking about ourselves and our encounters and misencounters with the communities with which we are engaged. Writing in the contact zone, we should find new ways of putting ourselves and others into text.

The Challenge of Doing Research in an Intercultural Context with a Transnational Team

It is challenging to develop such a broad and encompassing project. Not only a substantial element of interculturality is involved but also of transnationality, as the team gathers researchers from Brazil, Colombia, Namibia, Netherlands, U.S., Italy and Singapore. To engage in such a dialogue across cultures and nationalities, we feel that fostering fruitful work relationships has much to do with fostering good interpersonal relationships. This requires not only managing how people work together, but also dealing with how they come together as entire subjects, full of subtleties and complexities. If we do not understand each other's prior knowledge and conceptions, practices, interests, values, we cannot do the teamwork needed to carry out the demanding tasks in the field, ranging from working with teachers in local schools to engaging with local authorities and communities, from performing ethnographic/anthropological research to doing ecological studies. Moreover, how to balance team spirit and individual engagement is also a key point, because each team member also has his or her own project, under his or her own primary responsibility.

Interestingly enough, theory of mind is a key asset to be successful in coordinating our activities in such a project. By "theory of mind", we mean the ability to attribute mental states – beliefs, intents, desires, knowledge, etc. – to oneself, and to others, understanding that others have beliefs, desires, intentions, and perspectives that are different from one's own (Premack & Woodruff, 1978). To successfully coordinate our capacities and tasks in the lab and in the field, we need to strive for understanding others' minds, that is, doing proper folk psychology – as conceived here, exerting the cognitive capacity to explain and predict other people's behaviors and mental states (Ravenscroft, 2016) – is also instrumental to collaborative work in science, especially if we are quite different from each other, as it happens when we cross cultural and national frontiers. A momentous problem, then, is that we are often wrong when doing folk psychology. Minds are private, first person subjective experiences, and we should guess what others believe, want, know, etc. from what they tell us and how they act. And, as we all know, human communication is far from being transparent. We are often wrong in assuming what others are thinking from what they do and say.

This means that communication difficulties will happen all the time, and as a research team engages in transdisciplinary work, that is, recognize other partners, outside the academic world, as truly participating in research and innovation, communication difficulties may become even larger. What to do then? Our experience tells us that it is good to build trustworthy relations that allow us to be in the position of frequently asking ourselves, if we understood correctly what others in the team feel, believe, know, want. When trust is in place, it is easier to engage, also, with the art of saying things over and over again when they are important, and foster talking to each other within the team, engaging ourselves in the important (but sometimes difficult) art of conversation. Certainly, there are frictions and misencounters when a large team of researchers, from different national backgrounds, work together and, more than that, often share the same space in a household in a local community. The same can be said of a large team of researchers located in the intercultural contexts where we work. Tensions will happen, often due to miscommunication. When all the partners trust each other, they are reflexive enough and engage in open communication, it is not the case that friction, misencounters, tensions will not emerge; rather, what is important is that, when they appear, they are reflected upon and talked through, so that trustworthy relations remain. These are, in our view, key requisites for successful transnational and intercultural research and innovation.

Encounters and Misencounters in the Contact Zone

Nevertheless, it is evident that such a transnational, transdisciplinary and intercultural effort, gathering a number of different stakeholders, leaves room for both encounter and misencounter. In this section, we will consider examples from our field experiences.

A remarkable meeting in May 2017 offers an example of an encounter in the contact zone between university and community. From previous experiences, we are keenly aware of the importance of contributing to the conservation of fishing knowledge, especially in relation to identity construction processes. It is always touching to think that the knowledge of the elderly in the communities can be forgotten, and sometimes the only place where their knowledge is documented is in interviews, recordings and transcripts from a research team. For instance, in Diego Valderrama-Pérez's thesis (2016), the ages of the interviewed fishermen in the village of Taganga ranged from 69 to 83 years old, with ca. 75 years average. Among the eight interviewed fishermen, two had already died when the thesis was defended. Quite probably, the only place where their knowledge is documented in a systematic manner is in Diego's thesis.

When we talked about preserving fishing knowledge present in our field material, the idea of a museum naturally came to our minds. We concluded, however, that it was too early to say anything in this direction for the community. We didn't know whether and when the conditions necessary to develop the proposal might be in place. But in a meeting in May 2017 Lia, who both teaches at the School Sagrada Família and is a recognized leader in the community, told us she had a desire of creating a museum of fishing culture in Siribinha, in the headquarters of the local resident's association, then under her administration. There was a clear convergence between her wishes, as a leader in the village, and our ideas on preserving traditional knowledge. This was a key encounter because we established a partnership to develop a project of an ecomuseum and look for funding.

There were, however, some misencounters in the process, since it took a long time for us to find a museum scientist to join the team. When we finally contacted one who was interested in developing the project, Sidélia Teixeira, from the Faculty of Philosophy and Human Sciences at our university, Lia had grown impatient with the several months waiting for some progress. To keep the story short, after some rounds of conversations, we manage to get back to the initial agreement and we are now preparing a project of an ecomuseum. It became much larger than the original intention, since we are currently collaborating with local authorities (the Mayor and his team), and also mediating their relations with the communities so as to keep room for their participation in decision-making and management, in developing a plan for both environmental and knowledge conservation in the estuarine communities. We are writing at the same moment the process is developing under our noses, but from where we see it now, we may end up with a proposal combining a sustainable use conservation area, under the management of the communities (and, thus, conservation will be community-based), and an ecomuseum, where a convergence can emerge among environmental conservation, traditional knowledge valuing, and income generation through multifunctional tourism (that is, an approach combining nature, birdwatching, cultural, scientific, and beach tourism). We already developed ecotourism packages for local fishermen to operate, based on findings from our studies on the environments and birds in the estuary, and we are training them to become birdwatching guides while preserving local knowledge on birds.

It seems now that the encounter between university, community, and (currently) local authorities, in the contact zone in the estuary, is leading to new configurations of trust and new arrangements of power, in which resistance may mean integration into transcultural and transdisciplinary processes. These arrangements seem especially propitious in the current situation in Brazil, where a new extreme right-wing federal government will move in the opposite direction to environmental and traditional knowledge conservation. If in the past to go to the federal sphere of decision was a good strategy to assure conservation, now we need to get support from more local arrangements of power, like those at municipalities. In order for the fishing communities in the estuary of the Itapicuru to survive, they need to resist by integrating into these new arrangements of power, and the university can play a key role in mediating the multi-stakeholder processes needed for putting these arrangements in place, thus contributing for a sustainable transition in the communities (Scholz, 2017). Surely, we are writing as friction takes place in the encounters across difference we are experiencing (Tsing, 2005). We cannot be sure about the future, but at least we can say that the picture seems promising from where we stand now.

The ecomuseum is being conceived as a space of democratic exercise, a dialogical space of citizenship, a laboratory of human beings' experiences with nature, time, space, incorporating the concepts of the integral museum (Cerávolo, 2004). The proposal is that the communities actively participate in the construction of the structure and practices of the ecomuseum, engaging in participatory inventories, semistructured interviews, identification of possible liaisons, trajectories and dialogues about territory and history, new ecotouristic arrangements in the local environments, and so forth. If we manage to implement a collective process of musealizing cultural heritage and memory, more layers of intercultural dialogue will be added to the already multilayered process that we experience, as glocademics, in the local communities.

But, in order to show that not only encounters take place in the contact zone, it is also important to look also at examples of misencounters. A case in point is found in the reiterated need to explain the research and innovation goals to the communities. What we are doing there is alien enough for them to capture it only in bits and pieces. On one occasion, for instance, one of our team members, Diego, attended a party in Siribinha during his immersion in the community. A fisherman told him, then, that the project had to use its resources to teach villagers (especially, the elderly) how to read and write. This is but one of several examples of occasions where the community members showed misunderstanding about our goals. Despite the fact that we should be open to turn the communities' wishes into research goals, as one of the principles guiding our work, we can only be of some help when we have the required skills and expertise. Concerning the proposed task, we did not have such skills or expertise, and, besides, our involvement with several other community goals made it hard to add one more. The solution for such misencounters is to take every opportunity we have to explain to the communities our goals. We have thus taken the opportunities given by the culminating events of the school initiatives, at the end of the year, and the participatory planning workshops we carry out with the communities every two months to provide further explanations.

As an example, we can mention the culminating event in the Siribinha school in December 2017, which attracted the attention of many villagers. In the event we talked about the project; the students read a text on the history of Siribinha based on their school works; they exhibited a play entitled "pulling the net" (*puxada de rede*), showing a day in a fisherman's life, written by one of the students along the pedagogical work in the school; the pictures taken by students, teachers, fishermen, fisherwomen, and university students in a socioenvironmental perception activity were exhibited; short reports were made by the students on the educational work along the year; a video showing the teachers' reports on the collaborative work was exhibited; and, finally, we presented an ethnographic amateur movie edited from our interviews and participant observations.

The ethnographic movie was particularly important for our relationship with the community and its understanding of what we are doing there. The story behind the movie is also illustrative of the encounter made concrete in its realization. When chatting with a local fisherman, Diego was told that they wanted to see the videos he had been shooting during participant observations of fishing. Diego said he would be happy to give them a copy of the videos. But the fisherman said they wanted Diego to show the videos to them. From this statement, we got the idea of editing a movie. We didn't anticipate, however, that doing so would be so important to our work in the village, since it allowed the community to grasp in a deeper way what we mean by valuing and documenting its wealth of knowledge. These abstract words became much more concrete when they saw themselves through our ethnographic eyes.

A more complex story of encounter and misencounter marked our relationship with Pedro and Zefinha, traditional experts with a lot of knowledge and experience, who offered some of our most precious hours of interviews, about their practices of fishing in the mangroves around Siribinha. Their interviews have the exquisite touch that they talk with us together, complementing each other's talk, or sometimes speaking in counterpoint, as if commenting on one another's speeches. We had several beautiful encounters with them, who are among the poorest villagers, despite their wealth of knowledge. Their lives are quite hard, and from the toughness of their lives they succumbed to alcohol. Around a year ago, our communication became much more difficult and the spell of our encounter vanished. Since then our meetings, mostly in front of their house, seemed more like tortuous attempts to reach an encounter. We have been striving to move back to mutual connection, but it turned harder and harder as their health problems continuously increased. Now we can say it is just sad to talk to them, and we cannot get anymore not even a glimpse of their exquisite joint discourse.

Stories marked by engagement and encounter have also characterized our field studies on the local fauna. Zé Preto, one of the local fishermen, has been of great help in showing our students the knowledge and practices related to fishing, but also became a partner in field studies on a critically endangered monkey found in the estuary (Buff-headed capuchin, *Sapajus xanthosternos*, locally called "Macaco prego"). Zé Preto's classes on the pitfall called "Camboa" (or "Gamboa"), widely used in the Brazilian shore in the past, but now in decline (Ott, 1944; Pacheco, 2006) due to its cost and work demand, are a testimony of his engaging qualities. The same can be said of his knowledge on the monkeys, shared along the field expeditions. In turn, we have developed classroom innovations for teaching the schoolchildren about this endangered monkey species, which does not have a high reputation in the village, as it competes with the community for resources and is sometimes aggressive, especially towards fisherwomen working in the mangrove.

Nego (Mário Sérgio), our most frequent companion in bird surveys, offers another nice example of an encounter. Keenly understanding what we are doing and truly interested in our strategies, like using playback to attract birds, he has offered us nice interviews on local birds. For instance, he taught us how a near threatened hawk species from the mangroves (Rufous crab hawk, *Buteogallus aequinoctialis*, locally named Gacici) interact with fishermen, who pay attention to its calls to know that the tide is turning, and it is time to retrieve the fishes captured by their nets: a local saying states, "Gacici sang, the tide turned", a free translation from the Portuguese "Gacici cantou, a maré vazou". His ability to look for birds and perfectly arrange the boat for photographing, besides his enthusiasm when the hawk is found, has been enticing to birdwatchers who have been visiting the area since we reported the abundant presence of the rare hawk, among other rare species we also discovered in the Itapicuri River estuarine ecosystems. Currently, local engagement with our bird studies reached another level, as we are working with Nego and four other fishermen (Mauricio – Nego's son, Galego, Souza and Laelson) in developing ecotourism packages to be offered to visitors and we are also training them to guide birdwatchers.

The commitment of the teachers and, also, several other members of the communities to the research and innovation proposals, in a relatively short time, prompts us to reflect on the reasons for this positive answer to the request to access their knowledge and practices. Among the important factors in the construction of such trustful relationships, we identify our ethical concern with establishing a respectful relation with the teachers and traditional experts, explicitly treating them as keepers of their traditions and wisdom, and at the same time agents of innovations in their practices.

We consider that Siribinha and Poças, as traditional communities, can mark their difference through a performative cultural attitude (Silva, 2000), focused not so much on what they are in cultural terms, but, mostly, in what they can become, by a disposition to engage with the transformations they are likely to suffer in a way that keeps their identity. No doubt their cultural and historical identities will be transformed by ongoing processes that inevitable affect them, such as, say, the development of tourism in the north shore of Bahia. To keep their identity, resistance will mean certainly to engage and get integrated in these changes in a way that preserve their way of living albeit transformed, through a properly designed project of conservation and local development that shifts their course of transformation from the usual trend in the region, which leads to touristic development that displaces local communities or at least significantly hamper their way of living. It is possible to seek an alternative form of tourism from the one currently occurring in the estuary, with high socioenvironmental impact and limited financial return, and from the one usually implemented in the north shore. Nature, birdwatching, culture, scientific tourisms constitute possible avenues for the communities, and the relationships established among the university, local authorities and communities can be instrumental in developing and implementing a better plan for their future. In this context, whatever decision they may take about their future, it will tend to be more successful, if supported by the valorization of cultural legacy by old and new generations.

However, we cannot have the illusion that the relations within the communities are not marked by diverging interests. Local cultures comprise complex scenarios of alliances, competition and latent or explicit conflicts, which interfere in the way they represent themselves and lend themselves to representation. Thus, the project team has been carefully avoiding misencounters that may follow from neglecting power relations and conflicting interests pervading the communities, often motivated by different representations of the local identity. As a case in point, community members who wish to keep their traditions consider that asphalting the road that give access to the villages may represent a threat. But those who desire what they conceive to be progress and modernity consider this would represent an advance to the community. Socially engaged researchers cannot take sides in relation to these identity representations, which might foster local opposition and resistance. But they can play a role in creating spaces for broadening the awareness that identity representations under dispute are cultural productions, which can be always subjected to questioning and negotiation in view of community welfare.

Language Issues in Ethnographic Work

That language can be tricky was a lesson we were reminded of when trying to understand what the fishermen or fisherwomen were sometimes telling us. Even though we are all Portuguese speaking people, the specific language of the place can be rather demanding at times. The best example lies in our struggle to figure out what the word "quitanda" means in the local language (and elsewhere in the region).

They use the word "quitanda" to refer to one of the fundamental materials to do pitfalls like the "Camboa" (Ott, 1944; Pacheco, 2006), which is disappearing in the region, and the "Covo", still widely used (e.g., Magalhães, Costa-Neto & Schiavetti, 2011). Different people seemed to use the word, however, with different meanings. Sometimes it seemed to mean the wood splints used to assemble the main body of the pitfalls (Figure 6.1). Sometimes it apparently meant the leafstalk of palm trees like Piaçava (*Attalea funifera*) and Coconut (*Cocos nucifera*). Sometimes it seemed to refer to the palm Piaçava itself.

We could get a better glimpse of its meaning in an interview in which the fisherman Pedro described step by step how a Covo is made. He used the word "quitanda" sequentially to refer to the leafstalk of a palm tree,

Figure 6.1 Camboa (A) and Covo (B), pitfalls used in the Itapicuru River Estuary. A. Photo by Diego Valderrama-Pérez. B. Photo by Charbel El-Hani.

to the two halves of the stalk when divided, and to each wood splint resulting from breaking half of the stalk into smaller pieces.

However, we are still not sure about this meaning, because in subsequent interviews other fishermen and fisherwomen used the word in ways that did not fit easily with what we got from Pedro's interview. Albeit living in Siribinha for a long time, Pedro is originally from another fishing community in the region. His use of the word can vary, thus, from the more common way of employing it in Siribinha. We still need to check the meanings that appeared in that interview, despite the fact that they were so clear in Pedro's discourse.

Anyway, something noteworthy in the way he uses the word is that it may both refer generally to parts of plants (leafstalks of any palm tree) and get more specific meanings along the process of preparing the material to manufacture the pitfalls, moving step by step through the construction process.

Concluding Remarks

As Tsing (2005, p. 3) argues, far from being isolated cultures, traditional cultures have been shaped by national and transnational dialogues. They are not exemplars of "the self-generating nature of culture itself"; rather, "all human cultures are shaped and transformed in long histories of regional-to-global networks of power, trade, and meaning" (p. 3). Cultures are continually co-produced in interactions she calls "friction": "the awkward, unequal, unstable, and creative qualities of interconnection across difference" (p. 4).

It is all too easy to forget, when we come to the fishing villages, that the culture we are engaging with is itself a hybrid. It was in Bahia that Portuguese sailors first landed in the territories that would become Brazil, and the native Americans they met were quite apt in fishing. That was a colonial encounter between two fishing traditions, surely in quite unequal power relations. As a consequence, today we find in Brazilian fishing villages a hybrid between the Portuguese and Native American fishing cultures, with smaller input from Africans who were forced to come to Brazil as slaves (Ott, 1944). Even though it may not be so transparent nowadays, fishing culture in Brazil is itself the product of a transcultural dialogue, shaped by networks of power, trade, and meaning.

The very idea of proposing a dialogue between scientific and traditional knowledge may seem misguided by those who have been enticed by the "science wars", the debate over whether science is a privileged form of truth or just another political imposition. But we agree with Tsing (2005, p. 13) that it is more interesting to explore how knowledge moves, and how collaborations underlying knowledge production and maintenance take place. This seems better than simply assuming a Manichean for or against science positioning. We have not been, thus, much affected in our intentions by fearing science as an imposed knowledge. Rather, we have been much more self-aware and interested in how scientific and traditional knowledge move together and get into friction in the contact zone.

We are now also part of the processes that shape the villagers' histories, while we are at the same time shaped by them. The villages are places of interactions that can harbor encounters and misencounters, trust and mistrust relationships, and to think ontologically, epistemologically and ethically about how our work develops, we should focus on the interactive, mutually shaping nature of the relations we are building. We shouldn't lose "friction" (Tsing, 2005) from sight: what are the awkward, unequal, unstable, and creative qualities of our and their interconnections across differences? What qualities emerge when university and community, researcher and villager meet? How can and do mutually enriching relations and emergent cultural forms of research and innovation emerge from such interconnections? Can a community of practice indeed emerge in the collaborations between researchers and teachers, or between university and community? What are the risks and benefits ensuing from the relations and interconnections being built? How do different types of trust interact in these interconnections? As Tsing (2005, p. 13) writes, "Collaborations create new interests and identities, but not to everyone's benefit". New interests and identities have been created along our collaborative ways, but how are they seen from the eyes of the local teachers, of other community stakeholders, or of the junior researchers in the team? To nourish a reflective attitude towards encounters and misencounters across difference in the contact zone will continue to be crucial for the future of the mutual engagement between university and community in the fishing villages of Siribinha and Poças, in the north shore of Bahia, Brazil. In particular, this reflective attitude is much needed when we are not only doing research and developing innovations in an intercultural setting, but also engaging a transnational team gathering researchers from several disciplines, and, more than that, also other stakeholders, striving for building integrative knowledge in a transdisciplinary fashion.

Acknowledgments

We are deeply indebted to all the members of the communities of Siribinha and Poças who have shared their knowledge, time, experiences and affections with us since October 2016. In particular, we are thankful to the teachers of the schools Sagrada Família and Brazilina Eugênia de Oliveira for their continuous engagement and collaboration in the educational work in the villages. We should thank all the senior and junior members of our research team, who have been so enthusiastic and hard-working along the project, both in the lab and in fieldwork. We are indebted to research funding from the Brazilian National Council for Scientific and Technological Development (CNPq, grant n. 465767/2014-1) and the Coordination for Improvement of Higher Educational Personnel (CAPES, grant n. 88887.136397/2017-00) to the National Institute of Science & Technology in Interdisciplinary and Transdisciplinary Studies in Ecology and Evolution (INCT IN-TREE), where the project is included, and from CNPQ specifically for the project in traditional communities (grant ns. 303011/2017-3 and 423948/2018-0).

Notes

1. On the role of indigenous knowledge and ethnoscientific approaches in conservation and sustainable management of natural resources, see, e.g., Pierotti & Wildcat (2000), Rist & Dahdouh-Guebas (2006), Lynch, Fell & McIntyre-Tamwoy (2010), among several other sources.
2. Sensitive species only survive within a narrow range of environmental conditions and their disappearance from an area is an index of pollution or other environmental change.
3. By mentioning "cultural conservation", we do not hint at any idea of traditional culture as a legacy that cannot be allowed to change. Changing is an inherent part of culture. The key point is to strive for the conservation of those cultural processes that keep fishermen's and fisherwomen's identity.
4. The distinction we make between ways of knowing and forms of knowledge refers, as the designations indicate, to the process of knowledge production (*ways of knowing*) and the knowledge produced (*forms of knowledge*). This distinction is introduced because, when we consider traditional communities and scientific communities, there may be similarities and differences between both processes used to build knowledge, and the products of those processes, the forms of knowledge resulting from them. Given the distinctive nature of these two aspects, one being process the other product, we consider appropriate to take as a starting point, at least *ex hypothesi*, possibly distinct similarities and differences among them.
5. As examples, consider, for instance, the gap between biomedical knowledge produced by academic researchers and clinical and other medical practices (e.g., Bero et al., 1998), or between the selection of priority areas for biological conservation and knowledge produced by conservation biology (e.g., Knight et al., 2008), or between educational research and teaching practices (e.g., McIntyre, 2005), among many other possible cases in point.
6. The Mebêngôkre/Kayapó language belongs to the Jê linguistic family, while the Jul'hoan and !Kung are part of the !Kung dialect continuum, spoken in northeastern Namibia and the Northwest District of Botswana.
7. The term "misencounter", although not currently included in dictionaries, has been increasingly used in the academic literature lately. We decided to use it due to its rich meaning in relation to our fieldwork and theoretical efforts. We derived it from one of the works inspiring the study reported here, Toomey (2016).

References

Bero, L. A., Grilli, R., Grimshaw, J. M., Harvey, E., Oxman, A. D. & Thomson, M. A. (1998). Getting research findings into practice: Closing the gap between research and practice: An overview of systematic reviews of interventions to promote the implementation of research findings. *BMJ: British Medical Journal* 317: 465–468.

Bertuol-Garcia, D., Morsello, C., El-Hani, C. N. & Pardini, R. (2018). A conceptual framework for understanding the perspectives on the causes of the science-practice gap in ecology and conservation. *Biological Reviews* 93: 1032–1055.
Bertuol-Garcia, D., Morsello, C., El-Hani, C. N. & Pardini, R. (2020). Shared ways of thinking in Brazil about the science-practice interface in ecology and conservation. *Conservation Biology*, 34: 449–461.
Broome, B. J. (1991). Building shared meaning: Implications of a relational approach to empathy for teaching intercultural communication. *Communication Education* 40: 235–249.
Cerávolo, S. M. (2004). Delineamentos para uma teoria da Museologia. *Anais do Museu Paulista* 12: 237–268.
El-Hani, C. N. & Greca, I. M. (2011). Participação em uma comunidade virtual de prática desenhada como meio de diminuir a lacuna pesquisa-prática na educação em biologia. *Ciência e Educação* 17: 579–601.
El-Hani, C. N. & Greca, I. M. (2013). ComPratica: A virtual community of practice for promoting biology teachers' professional development in Brazil. *Research in Science Education* 43: 1327–1359.
El-Hani, C. N. & Mortimer, E. (2007). Multicultural education, pragmatism, and the goals of science teaching. *Cultural Studies of Science Education* 2: 657–702.
El-Hani, C. N. Silva-Filho, W. J. & Mortimer, E. F. (2014). The epistemological grounds of the conceptual profile theory. In: Mortimer, E. F. & El-Hani, C. N. (Eds.). *Conceptual Profiles: A Theory of Teaching and Learning Scientific Concepts* (pp. 35–65). Dordrecht: Springer.
Freire, P. (1970/1987). *Pedagogia do oprimido*. São Paulo: Paz e Terra.
Gavin, M. C., McCarter, J., Mead, A, Berkes, F., Stepp, J. R., Peterson, D. & Tang, R. (2015). Defining biocultural approaches to conservation. *Trends in Ecology & Evolution* 30: 140–145.
Haraway, D. J. (2008). *When Species Meet*. Minneapolis, MN: University of Minnesota Press.
Knight, A. T., Cowling, R. M., Rouget, M., Balmford, A., Lombard, A. T. & Campbell, B. M. (2008). Knowing but not doing: selecting priority conservation areas and the research-implementation gap. *Conservation Biology* 22: 610–617.
Lave, J. & Wenger, E. (1991). *Situated Learning: Legitimate Peripheral Participation*. New York, NY: Cambridge University Press.
Ludwig, D. (2016). Overlapping ontologies and Indigenous knowledge. From integration to ontological self-determination. *Studies in History and Philosophy of Science* 59: 36–45.
Lynch, A. J. J., Fell, D. G. & McIntyre-Tamwoy, S. (2010). Incorporating indigenous values with 'Western' conservation values in sustainable biodiversity management. *Australasian Journal of Environmental Management* 17: 244–255.
Magalhães, H. F., Costa-Neto, E. M. & Schiavetti, A. (2011). Saberes pesqueiros relacionados à coleta de siris e caranguejos (Decapoda: Brachyura) no município de Conde, Estado da Bahia. *Biota Neotropica* 11: 45–54.
McIntyre, D. (2005). Bridging the gap between research and practice. *Cambridge Journal of Education* 35: 357–382.
Ott, C. F. (1944). Os elementos culturais da pescaria baiana. *Boletim do Museu Nacional* 4: 1–67.

Pacheco, R. S. (2006). *Aspectos da Ecologia de Pescadores Residentes na Península de Maraú-BA: Pesca, Uso de Recursos Marinhos e Dieta*. Brasília, DF: Graduate Studies Program in Ecology, University of Brasília (UNB), Masters Dissertation.

Pain, R., Kesby, M. & Askins, K. (2011). Geographies of impact: power, participation and potential. *Area* 43: 183–188.

Pardini, R.; Rocha, P. L. B.; El-Hani, C. N. & Pardini, F. (2013). Challenges and opportunities for bridging the research-implementation gap in ecological science and management in Brazil. In: Raven, P.; Sodhi, N. S. & Gibson, L. (Eds.). *Conservation Biology: Voices from the Tropics* (pp. 75–85). Oxford, UK: John Wiley & Sons.

Pierotti, R. & Wildcat, D. (2000). Traditional ecological knowledge: The third alternative. *Ecological Applications* 10: 1333–1340.

Pratt, M. L. (1992). *Imperial Eyes: Travel Writing and Transculturation*. London: Routledge.

Premack, D. & Woodruff, G. (1978). Does the chimpanzee have a theory of mind? *The Behavioral and Brain Sciences* 4: 515–526.

Ravenscroft, I. (2016). Folk psychology as a theory. In: Zalta, E. N. (Ed.). *Stanford Encyclopedia of Philosophy* (Fall 2016 Edition). Available at https://plato.stanford.edu/archives/fall2016/entries/folkpsych-theory/, accessed December 28th 2018.

Rist, S. & Dahdouh-Guebas, F. (2006). Ethnosciences—A step towards the integration of scientific and indigenous forms of knowledge in the management of natural resources for the future. *Environment, Development and Sustainability* 8: 467–493.

Scholz, R. W. (2017). The normative dimension in transdisciplinarity, transition management, and transformation sciences: New roles of science and universities in sustainable transitioning. *Sustainability* 9: 991.

Silva, T. T. (2000). A produção social da identidade e da diferença. In: Silva, T. T. (Org.). *Identidade e Diferença: A Perspectiva dos Estudos Culturais* (p. 73–102). Petrópolis: Vozes.

Sousa Santos, B. & Meneses, M. P. (Orgs.). (2009). *Epistemologias do Sul*. Coimbra: Almedina/CES.

Sundberg, J. (2006). Conservation encounters: transculturation in the 'contact zones' of empire. *Cultural Geographies* 13: 239–265.

Toomey, A. H. (2016). What happens at the gap between knowledge and practice? Spaces of encounter and misencounter between environmental scientists and local people. *Ecology and Society* 21: 28.

Tsing, A. L. (2005). *Friction: An Ethnography of Global Connection*. Princeton, NJ: Princeton University Press.

Valderrama-Pérez, D. F. (2016). *Diálogo entre conhecimentos científicos escolares e tradicionais em aulas de ciências naturais: Intervenção e pesquisa na comunidade de Taganga (Magdalena-Colômbia)*. Salvador, BA: Graduate Studies Program in History, Philosophy and Science Teaching, Federal University of Bahia (UFBA)/State University of Feira de Santana (UEFS), Ph.D. Thesis.

Viveiros de Castro, E. (2009). *Métaphysiques Cannibales*. Paris: Presses Universitaires de France.

Wenger, E. (1998). *Communities of Practice: Learning, Meaning, and Identity*. New York, NY: Cambridge University Press.

7 The EC-MSC Glocademics project at the Universidade Federal da Bahia

Interview with Manuela Guilherme

Sávio Siqueira

In Memory of Professor Suzana Cardoso, Projecto ALiB, Instituto de Letras, Universidade Federal da Bahia, Salvador, Brazil

Manuela Guilherme, senior researcher at the *Centro de Estudos Sociais* (CES), *Universidade de Coimbra*, Portugal, has always been an inquisitive and prolific scholar. Born in Mozambique to Portuguese parents, Manuela moved to Lisbon when she was 17 to further her interest in languages and cultures at the University of Lisbon where she studied English and German languages and literatures, besides Portuguese and French literature. At home, as an infant, she had learned Portuguese and two African languages and later, at school, German, French, and English, while in Mozambique. Along her successful academic career, which includes a PhD degree from Durham University, UK, Manuela Guilherme led different projects funded by the European Commission, including ICOPROMO – Intercultural Competence for Professional Mobility (Leonardo da Vinci Program), also sponsored by the European Centre for Modern Languages, Council of Europe and INTERACT – The Intercultural Dimension of Citizenship Education (VI Framework Program), also sponsored by the Gulbenkian Foundation. Both aimed at professional development of teachers, researchers, non-governmental and governmental workers dealing with cultural diversity. Her research work in Latin America started in 2011, as the scientific co-coordinator of the RIAIPE 3 project, funded by the European Commission (ALFA Program), focusing on the role of higher education in equity and social cohesion carried out by scholars and researchers from 22 Latin American and 8 European higher education institutions (see chapter 2).

As recipient of a Marie Sklodowska-Curie research grant (2014–2017), her latest project, GLOCADEMICS – "'Glocal Languages' and 'Intercultural Responsibility' in a postcolonial global academic world: Power relations between languages/cultures within and between research groups", was developed between the CES, its leading host institution, and

DOI: 10.4324/9781003225812-8

the Department of Modern Languages of the *Universidade de São Paulo* (USP), the host institution in Brazil. The project included research carried out also at the *Universidade Federal da Bahia* (UFBA), *Universidade Federal do Sul da Bahia* (UFSB) and at *Universidade Federal do Paraná* (UFPR). As a groundbreaking endeavor, involving Brazilian research teams that developed transnational research and with a focus on their exchange practices with European partners, Manuela Guilherme justifies "Glocademics" by arguing that plurilingualism and intercultural epistemological exchanges in transnational research deserve particular scientific attention. The findings brought about at this point by this study are expected to raise interest, however, this interview is going to be specific about the development of this two-year research project at the *Universidade Federal da Bahia*, through a conversation between the Principal Investigator, Manuela Guilherme, and Sávio Siqueira, a professor of English and Applied Linguistics at this university.

GLOCADEMICS involved working with five research groups in three of the above mentioned universities in Brazil (USP, UFBA and UFSB), through individual meetings, document reading and analysis, meeting observation, and finally focus groups interviews. The two research groups from Bahia Federal University (UFBA), in Salvador, Bahia, Brazil, were the ALiB project team (*Atlas Linguístico do Brasil* – Linguistic Atlas of Brazil), based at the *Instituto de* Letras - Institute of Letters, in cooperation with other departments, and the LEFHBio research group (*Laboratório de Ensino, Filosofia e História da Biologia* – History, Philosophy and Biology Teaching Laboratory), based at UFBA's Institute of Biology. This interview will focus mainly on the work that Manuela Guilherme conducted with these two research groups in order to find out about their management of linguistic resources and their attitude toward intercultural epistemological collaboration throughout their experience of trans- and intra-national research.

However, prior to this second phase of the project and as a preparation for it, Manuela Guilherme also undertook a study with (9) colleagues from UFBA's Institute of Letters, as well as at USP (Universidade de S. Paulo) and UFPR (Universidade Federal do Paraná), a total of 27 professors who teach Portuguese, both as mother tongue and as foreign language, English and Spanish as additional languages and Indigenous languages, the latter only at the University of S. Paulo.

Some information about the two projects investigated in UFBA will be important at this point. The ALiB Project (see chapter 4), a nationwide prestigious linguistic enterprise which has been led for several decades by the UFBA team and has published, among several other works, the *Brazilian Linguistic Atlas* in two volumes (Cardoso et al., UEL Editora, 2014). The LEFHBio (see chapter 6) coordinates large transnational and transdisciplinary projects dealing with ecology, biology, conservation, philosophy of science and education. Amongst their largest is INCT IN-TREE project (2016–2022), addressing ecology and evolution, particularly engaging with connections between these sciences and mathematical and computational modeling, epistemology, ethics, and on the relationship between science, technology, and society studies. It gathers an international network of 31 Brazilian several universities, comprising 240 researchers, among them 45 researchers from 37 different institutions in Europe, North and South America, Asia, Africa, and Oceania. Another international project, started in 2017, on "Intercultural Education as Dialogue between ways of knowing and knowledge frameworks: Strategic and collaborative research in traditional communities". It is building a knowledge base, integrating academic scientific knowledge, about traditional knowledge from fishing and indigenous communities both in the northeast of Brazil and Namibia, and practical knowledge from environmental technicians and teachers, for use in conservation and education efforts. This project includes researchers from the Netherlands, Italy, Denmark, Australia, and the United States.

What Made You Expand the Glocademics Project to the Universidade Federal da Bahia?

Although the University of S. Paulo was the host institution of the Glocademics project in Brazil, besides the Center for Social Studies which was the overall project leading host institution, I had already included the University Federal of Bahia (UFBA) in my proposal to the European Commission. I made this decision because I was well acquainted with UFBA, mainly with colleagues at the Institute of Letters and at the School of Education, and I knew that I could count on their support, more particularly with yourself and dear colleagues Denise Scheyerl and Edleise Mendes who was in the meantime carrying out a related international postgraduate project in Coimbra, under my supervision. Besides, hospitality in Bahia is world famous and, above all, the Portuguese-African cultural mix, all of which help me feel more at home and which helps a researcher mix with other researchers more easily and, therefore, capture relevant underlying details. I had been an eager reader of Jorge Amado in my teens in Mozambique, and his books, settled in Bahia, were then understandably more appealing to me than Portuguese literature. Mia Couto, about my age, had not yet started to publish books ... and unfortunately there were no other Mozambican authors being published by then, at least that I knew of.

But there is more. I had been at your university in 2007, 7 years earlier, if you remember. By coincidence, I was about to travel to Santa Catarina, in the South of Brazil, when you contacted me about my 2002 book, which you were using for your doctoral thesis. So, I planned a visit to UFBA to give a seminar that you and Professor Scheyerl, your supervisor, organized, and I remember the great interest shown by your postgraduate students and the good discussions they raised about my research projects and publications on critical pedagogy, language education, and intercultural communication. They were by then well acquainted with issues of critical pedagogy, South-South and South-North epistemological relations and decoloniality. Much more than in my more recent experiences. The same has happened, unfortunately, almost everywhere else. Academic work in languages, cultures and interculturality has been regressing in their critical approaches, as I see it, all over the world, reflecting a shallow understanding of the role of languages in globalization. I was back to Salvador in 2012 and, again by sheer coincidence, for one of the Latin American meetings and complementary conference of the RIAIPE3 project, mentioned above, and which were held at UFBA. I had then one more opportunity to meet the colleagues from the Institute of Letters, actually, I remember that you were precisely launching a book for which you had kindly invited me to write the preface. This is only one example that shows that we had continued to collaborate since my first visit. Thus, it made sense that I did not leave UFBA outside the following project and, with this interview, I would like to express my special gratitude to my colleagues and friends at UFBA who gave such a warm and supportive welcome to Glocademics there.

What Made You Choose the ALiB Research Group for the Development of the Research Study Carried Out by the Glocademics Project?

The ALiB Project's main goal has been to produce a Linguistic Atlas of the Brazilian Portuguese language on a national basis. Given that the Brazilian territory is immense, covering around 47% of the whole area of South America, this project has been a tremendous challenge and a reference for linguists around the world involved with linguistic atlases. I remember them mentioning the interest raised among scholars from the United States and Latin America, Cuba, and Europe, more particularly from Portugal and Spain (Galicia), Germany, Italy, etc. Moreover, the interdisciplinary character of their study, its methodology, the amount of data collected and the accuracy of data analysis, both sociological and linguistic, with a significant technological input, has also been worth of recognition abroad, mainly in Europe and Spanish-speaking Latin America. The project's large bulk of outcomes is much more recent than the beginning of data collection which, to be more precise, has been completed over a decade ago and comprehends publications, workshops, and the organization of conferences as well as the participation by their research members in events worldwide, China included. I also learned that the ALiB Project has been grounded in a long tradition in Brazil and more particularly in Bahia where such studies, for example about the particular dialects in Bahia, started in the 1960s. Its coordinators, the late and dearest Professor Suzana Cardoso and Professor Jacyra Mota, as well as the whole ALiB research group at the UFBA, were extremely welcoming to the Glocademics project. This is not so usual as one might think since research groups in general, although surprisingly, do not easily accept being themselves the subject of research, more particularly when this is focused on topics that are not part of their expertise but are nevertheless part of their work routines, such as language and cultures which was not the case with the ALiB team.

In addition, I perceived that the ALiB project could provide a solid basis for the development of the concept that I had introduced and was developing, that of "glocal languages", since they have collected accurate data about the varieties of Brazilian Portuguese language throughout the many and large national states. The ALiB project was not meant to collect data about the other languages in Brazilian soil, neither indigenous nor immigrant languages, but it aimed at collecting data about their influence in the semantics and phonetics of Brazilian Portuguese. The Glocademics project's theoretical framework relies on the development of a three-axis matrix for pluringual and intercultural epistemological exchange that encompasses the concept of (a) "glocal languages" (Guilherme 2018, 2019; (b) "glocademics/glocademia" (see Introduction) and (c) "intercultural responsibility" (see chapter 8, Guilherme 2020a and 2020b). My elaboration on Portuguese as a glocal language, not only across the world but also in such a wide continental region, as well as the impact of its local use, both synchronically and diachronically, as in the "glocalization" of each other language, is solidly sustained by the knowledge put forward by the ALiB project. This is one good example of how important it is that the researcher keeps an open mind to different hypotheses ahead when s/he goes into the field.

Actually, my empirical research about Portuguese as "glocal language" was also part of the first phase at UFBA, through which I had long interviews with teachers and examined some of their teaching programs for undergraduates, including teachers of Portuguese. This new terminology and conceptual framework – "glocal language" – was embraced by UFBA's colleagues, not only during the project activities but with evidence given, for example, by an article entitled "Portuguese as a glocal language: Socio-historic and linguistic aspects of its conformity" (Machado Filho and Oliveira, 2017). Machado Filho participated actively in every step of the project at UFBA, in phase 1 as a teacher of the Portuguese language at the undergraduate level, in phase 2 as an ALiB member, at the final interdisciplinary event and as one of the coordinators of the post-graduate program to which I also contributed with a module.

And What about the LEFHBio? It Must Have Been More Difficult to Reach It and Even to Convince Them about the Relevance to Them of the Glocademics' Enterprise

I must say that the LEFHBio at UFBA was also very welcoming to my project throughout the process that lasted two years between the first contacts and the end of the project in Brazil. Professor Charbel El-Hani, the Lab coordinator, was recommended to me by another colleague of ours at UFBA's Department of Education with whom I had collaborated in my previous RIAIPE3 project, Professor Robinson Tenório. Understandably, they showed some perplexity about my request, to start with, but never closed doors, on the contrary, they were always very open and gradually more and more collaborative. I still keep in close contact with Professor El-Hani and I believe he has since then become more critically aware of the plurilingual and intercultural approaches that were already in place. By the way, he has just spent 18 months at CES, Coimbra, in order to develop his intercultural and transdisciplinary epistemological research approach from the perspective of the Epistemologies of the South and his research group remained as a partner in the project proposals that I have submitted to the European Commission. My collaboration *with* this research group entailed individual interviews with each member, meeting attendance and observation, careful reading of their publications and, thereafter, interpretation of oral and written data collected, that is, of both meetings and publications, in the light of the Glocademics' aims and objectives.

My serious involvement with their work, I believe, also convinced them about the relevance and positive effects that my work could have on theirs and *vice versa*. This was, it seems to me, an inspiring example of collaborative interdisciplinarity-in-*praxis*. I could talk informally to postdoctoral researchers, who were spending some temporary study stays there, from France, Colombia, Norway, South Africa, among others. These talks, which can be called informal and open interviews, were very helpful not only to make me more familiar with their work but also to build some mutual trust. Meeting observation and informal talks, simultaneous to my readings of their publications and reports, paved the way for the successful final group interview. My readings consisted of a selection of a book and several articles among those which were accessible to me in terms of content and which could raise any issues related to my own research questions and objectives. It was very rewarding not to discard such an interdisciplinary enterprise just because you have to venture into a knowledge field of which you are ignorant. In sum, it meant indeed a worthy adventure.

How Productive Were the Discussions toward the Objectives You Previously Established in the GLOCADEMICS?

The discussions were extremely productive in both groups. By discussions, I mean the initial individual interviews, fundamental for both parties to be acquainted with each individual projects and for me to find out where both projects, mine and theirs, could meet and be reciprocally helpful, and the final focus group interviews, one with each group. In the meantime, between the initial individual open interviews and the final focus group conversation, I had to make myself knowledgeable about the research groups' work and earn their interest and trust in my project by reading some of their publications and reports, besides attending some of their meetings. All the knowledge I had acquired about their research work, during this two-year study, about the difficulties they had found and the solutions to which they had come and experienced and created themselves, with regard to languages and cultures, was extremely helpful in order to design the final focus of each group interview guide. I prepared a different focus group discussion plan for each research group, which was checked beforehand by the respective group coordinator and shared with the other members. Therefore, they knew in advance what we were going to talk about so that they could prepare experiences to share and arguments to put forward.

These were challenging tasks to overcome but, in the end, they gave a good contribution to my project outcomes. The researchers in both groups generously agreed to participate in the focus groups, in the same way as they had done before throughout the whole project. They all showed up on time and only left in the end, the focus group session lasted for one hour, sharp. Not only did they generously offer their precious time for this purpose, but also identified the contributions this study might bring to both projects, mine and theirs. Through a meta-research process, both UFBA research groups became gradually more aware of the linguistic and intercultural issues raised in relation to their own research work, each research group from its own perspective. This was made evident in each step of the Glocademics project, as the participants demonstrated to have appropriated the subject matter and taken hold of the issues in question. Every member of both research groups was academically so hospitable as to take time to tell me about the history of their projects, different research stages, the anecdotes they had experienced during their field work, etc.

What Differences Could You Notice between the Work Developed with LEFHBio and the ALIB Group? How Similar/Different Were They?

Since knowledge about our world has been divided into the conventionally called disciplines, from this perspective, I must say that their work was totally different from each other. One research project was about linguistics, working on a linguistic atlas, the other was about ecology, conservation, and education, working on different issues ranging from ethnobiological studies on fishing communities to educational research on a diversity of classroom innovations. However, your question is very thought-provoking and unexpected because anyone could have taken for granted that they are divided by a disciplinary barrier and, therefore, had nothing in common. Nevertheless, both were interdisciplinary projects, both included education in their aims, especially teacher education, and both were simultaneously the subject of the same research project, Glocademics, with its own established aims, and, therefore, carrying out meta-research on their activities with regard to their plurilingual and intercultural epistemological activities within their own projects. The two research groups were sharing a similar circumstance, raising interrogations and reflections which were often alike, not knowing much about each other until they met in the final interdisciplinary event. I had mentioned to them, in the beginning, that I was also working with other groups, however, this was not an issue to dig into, for methodological reasons.

Above all, both projects, independently from their different disciplinary allocations, cherish Brazilian particularities and idiosyncrasies and value traditional and popular knowledge which resists to hegemonic globalization and they are both attempting to take a collaborative and mutual learning approach to that knowledge. This common feature to both projects, which I perceived as a good example of civic responsibility in research, provided me with fertile ground for the development of another concept which I had introduced before and was also developing, that of Intercultural Responsibility (see chapter 8). The crossing lines of interculturality in both projects are situated between the local and the global, in Bahia, across the national states of Brazil, and internationally, while promoting an emancipated ecology of knowledge at different levels of cultural diversity. Their expertise and the breadth and accuracy of data collection and analysis by each of these research groups, give account of the importance of their work.

What Did Participant Researchers Say Concerning the Position of Portuguese in the World Today and How Did They See the Current Hegemonic Position of English in the International Academic Scenario?

The ALIB researchers were very explicit about it. The same with the LEFHBio researchers for whom the use of English in science is even more pressing. Although these are very different research groups, one in sociolinguistics and the other in biology, both interdisciplinary, their research activities focus on field work in Brazil, in Portuguese, and they offer international publications and conference papers in English and French (more for the LEFHBio), besides the ones in Brazilian Portuguese as one should expect (more for the ALiB). I gathered that the international activity happens in a two-way direction for both projects but with different intensity. On the one hand, the ALIB international activity concentrates mainly on presenting the ALIB results internationally and by looking for European expertise through sending doctoral and post-doctoral researchers supported by university agreements. The ALiB researchers very clearly recognized the prevalence of European influence in their work, most often directly (from France, Spain, Portugal, and Italy), but also indirectly, for example from the Linguistic Atlas of Uruguay with German influence. Of course, they also receive much input from work developed in the United States mainly with regard to the use of computation and quantitative methods. However, their activity and production concentrates more inside national borders and activities outside borders are subsidiary contributions in terms of scientific relevance.

On the other hand, I understood that with the LEFHBio research group, besides close collaboration also with France, Italy, and Spain, particularly with the Basque country, their international collaboration is not only more diversified throughout the world – besides Latin America, Colombia in particular, with close cooperation with Scandinavian countries, Southern Africa and North America as well – but also more intensive in every aspect, that is, it is part of their routine. While the ALiB's research work lies on the development of a linguistic atlas of a very definite entity, a particular language within a particular nation-state, despite its intensive variability in both ways, for the LEFHBio research group and their peers, international comparative analyses play an important role in their activities. Besides, intensive diversity of all kind underlies a common language, English, which frames both terminology and conceptual framework, for every participant worldwide, although still asymmetrically. Their positions as Brazilian and Portuguese-speaking scholars have a negative impact with regard to opportunities for publication in international journals, in this case, regardless of their large-breadth work or important positions held in international associations and editorial boards.

On the whole, they were adamant in stating that English is so exclusively dominant in the international academy nowadays that it ends up determining, I would say limiting, the findings, topics and perspectives to be taken into account in science. And this is a very serious statement, supported by other natural sciences researchers. The valuable role played by the use of the English language worldwide in the internationalization of higher education in the last decades of the 20th century which, on the one hand, led to the strengthening of institutional democracy and intensive transnational research but, on the other hand, it became excessive at the turn of the century with linguistically and interculturally blind rankings. The regression from the critical cultural awareness approach to language education, proposed by Mike Byram et al., in the 1990s, at one end, to *lingua franca* by Jennifer Jenkins et al. in the 2010s, at the other end, did not help either and gave an illusory sense of easiness in language use and intercultural epistemological negotiation which has been pernicious to scientific research.

It is interesting to note that, in Brazilian universities, language departments scholars are more reactive to publish in other languages, in English more particularly, than scholars in the natural science departments, as shown by the national rankings. The participant researchers were also clear about the benefits and limitations that it imposes in research work, international collaboration and dissemination of results. On the positive side, it has enabled, within more limits than it is acknowledged by the English-speaking inner circle, worldwide scientific collaboration, whereas, on the negative side, the broad use of terminology in English has paradoxically reduced, even eliminated, conceptual diversity. The worst of all is not the conceptual limitations themselves, but the lack of awareness of what is simply erased, that is, the failure is not so much caused by the use of English but by the disregard to the process of "intercultural translation", the absence of critical intercultural awareness and the "lingua franca" fantasy. There is a limitation of knowledge validation and dissemination, established by the limits of the English language. I believe that I can make such a statement because I have access to several languages, in academic register, which enables me realize how much knowledge is thrown away with the bathwater… In my view, such blind imposition of monolingualism in the academy and in science and, even worse, the apparent simplicity of technical translation, both human and computing, narrow terminology due to conceptual compression, requires urgent attention. Not only does it leave conceptual misunderstanding unattended but it also restricts access to much epistemological wealth in the world, carried out in other languages and in other spaces. In the end, it reduces the scope of grounding evidence.

Every participant researcher reported in detail her/his experiences regarding conference papers, publications and research collaborations. It is common understanding that the ones who are not proficient in academic English are the only ones who end up being deprived of access to fundamental knowledge. On the contrary, the lack of voice and visibility awarded to knowledge in other languages reduces the scope of validity of knowledge produced by academics who are monolingual, either in English or in any other language. It should not be forgotten that there are languages covering such a large population that they can survive endogenously, both economically and epistemologically, and this can be said, for example, of Brazil. An important benefit brought by English, as a world language, has been, for example, its enormous potential for exogenous collaboration. However, once it comes to produce knowledge massively and in a closed epistemological circle, therefore apparently erasing meaningful difference, it loses its pertinence and ends up causing enraged opposition, and ultimately competition, to its dominant position.

What Responses Did You Get from Participant Researchers in Both Groups Concerning Possible Situations of Prejudice or Misconceptions on the Side of International Colleagues They Might Have had to Deal with? Was There Any Sort of these Reported and/or Discussed?

The idea of globalization has become economically bounded and territorially ignorant, that is, the relationship between the conceptualization of the global and territorial size is often reversed. This was expressed by both groups with examples of issues raised with regard to scientific work validation. The nationwide Linguistic Atlas of Brazil raises surprise and interest in international conference presentations, for example, for the huge area covered and its innovative methodology, only compared to the US Linguistic Atlas, according to worldwide recognized experts such as Labov, as mentioned by the ALiB researchers. According to them, the breadth of their study has turned comparative analyses with other linguistic atlases more difficult, mainly with those in Europe (Portugal, France) and in other Latin American countries (Uruguay, Colombia, Mexico) because of the different territorial size and smaller variability. This has caused some confusion with other colleagues who are not aware of the territorial extension of Brazil, its cultural diversity and wealth and of its researchers' outstanding expertise also resulting in high competition inside the Brazilian scientific community.

The LEFHBio researchers had more anecdotes to tell since they are internationally more active. For example, they mentioned they had often experienced difficulty in having publications in English approved about issues covering large areas in Brazil, for example about the *Caatinga*, a type of vegetation exclusive of Brazil's semi-arid Northeast, which covers a region that is larger than the totality of western Europe, under the argument that this study was about a local issue of reduced interest. For another article on a concrete period of the history of science in Brazil, the author was requested to include an introduction with a "brief" history of the development of science in the country, probably assumed to be of a reduced scope, on the grounds that the journal was not focused on Latin America and, therefore, its readers wouldn't be knowledgeable about science in Brazil. We may then conclude that this is more justifiable than the lack of knowledge about Europe or the United States, whose scientific development is assumed to be more evident and total ignorance about it unacceptable. Should any European author be asked to make a brief account of the history of science in Europe for an article introduction? The participants had a joke about the fact that a study about Chesapeake Bay, in the Atlantic coast of the United States, should be of global scientific value while one about the All-Saints Bay (*Todos-os-Santos*), in the Atlantic coast of Brazil, is of local relevance only. One is scientific knowledge and the other is exotic knowledge, I would say, no matter the inner accuracy of the study. Above all, this is, in my view, all about epistemological interculturality.

What Did the Members Say Concerning Difficulties and Advantages in the Exchange Work with Researchers from Other Linguistic and Cultural Contexts? More Particularly about Intercultural Mistakes/ Blunders throughout the Group's Interactions with Colleagues from Other Countries?

The members were very much aware of possible misunderstandings and of the unavoidable ambiguity both in the translation of concepts from one language to another or by the insertion of terminology in English in other linguistic contexts, despite the spread use of English in science and the normativity of scientific discourse in English. It was though unexpected to me to hear about the extent, of which they were themselves surprised, of unknown, or twisted, words and expressions in Portuguese, imported from several other languages, which they had not heard before and often could not even understand during their fieldwork. The ALiB researchers told many curious anecdotes during their fieldwork, mainly throughout interview transcriptions, which are difficult to me to translate into English. The LEFHBio researchers also reported having some difficulty in using specific traditional terms for tools used by the fishermen, for example, with whom they were carrying out their research activities. However, without surprise, while for the ALiB project the Brazilian Portuguese language was their research focus and the English language was subsidiary for international dissemination of their results, for the LEFHBio research group the English language was more integrative as most of their sources were in English and international publications and collaboration played a larger role in their academic careers.

Although both research groups used some technical words in English, without translating them, this was much more prominent with the biologists, both of Portuguese (Brazilian) or Spanish (Colombian) mother tongues than for the ALiB researchers, except for computational terminology. Understandably, the ALiB researchers, as sociolinguists, were more aware of ambiguities and mismatches emerging from language contact. Some of the LEFHBio participants mentioned how they felt while writing in each of the languages. One of them, for instance, noted that he was more precise in his English performance, since his semantic field was more restricted in this language while in Portuguese he could get lost in style and in the idiosyncrasies of the language. Another explained more or less the same process by saying that, in English, he followed the bibliographical sources more closely with regard to terminologies, while in French, his mother tongue, he felt more free to develop some critical analyses. It is noteworthy that they also pointed out that some colleagues only cited authors in the Portuguese language, however, they said, in this case, their work rested encapsulated in one body of literature and out of the international circles. As far as I see it, it is today difficult anywhere to accept that researchers, at an advanced academic level, say doctoral and post-doctoral, do not have, if only to some extent, multilingual capacities, whatever your mother tongue. Although you can have access to translations, the digging into original work in other languages always gives you an intercultural dimension to your field of knowledge that is inescapable in contemporary "Glocademia", my term, which certainly implies taking more risks.

Some of the LEFHBio researchers mentioned the risk of misunderstanding while translating concepts, say, from English into Portuguese. I would like to add that this risk is present even when you use the English word while the notion you have of it remains attached to your own native conceptual framework. This can become more complex when you adopt that term or expression in the context of a specific theoretical School or Theory, since you nevertheless make the transfer of the original concept to your argument within a particular context. Conceptual translation across plurilingual and intercultural epistemologies has been neglected. The participants gave precise examples, that is, for example about the use of the adjective "Darwinian" in English contexts, as its meaning has different connotations between English-speaking theoretical sources and Spanish-speaking ones. Another example was about the term "science teaching" which one of the participants used with a group of German scholars who had a different understanding of the term. However, the participants explained that the categories of diversity are not mutually exclusive, they can be complementary, and that, in this case, it was not only because they were all using English as a medium, none of them being a so-called native speaker or because they belonged to different national cultures, instead because they stood from different fields in biology. While the German group's field was placed in the Philosophy of Biology, the speaker (a LEFHBio member) was involved both in educational programs and on research, which is very common in Brazilian universities. This means that it is important to be aware of the multiple elements, categories and backgrounds that come across in an epistemological negotiation. The perception that meaning is straightforward just by using a common language is illusive.

What Were Their Responses toward Linguistic and Cultural Hegemony in Scientific Production? Were You Surprised by Any of Their Responses? Why? Why Not?

This is a very good question. To what extent can a researcher be surprised by the implementation of her research plan? To what extent can you allow yourself to be surprised, in order to take advantage of all the unexpected opportunities, without failing to fulfil the research plan previously approved? How strictly should you meet the objectives that convinced your evaluators to award you the funding? To what extent should the researcher, for ethical reasons, carry out her study *with* and not *on* the research subjects, that is, respect their interests, opinions, rhythms? Should you feel obliged to give back to the research participants? There are officially five elements that count most in the evaluation of a proposal submitted to the EU research funding programs, which sometimes may happen to be contradictory, and I imagine that they often are, in the assessment procedure: (a) your previous research experience; (b) the feasibility of your research plan; (c) innovation, that is, the groundbreaking nature of your research plan; (d) the interdisciplinary nature of your research plan; (e) the relevance of your study to your research field. Finally, the most decisive element of all is submitting your proposal to the most adequate panel, among the ones available. As you can imagine it is very difficult to balance every five elements between them, the elements above, and between each one of the panel members and the importance each element is awarded by each of the evaluators' own different backgrounds.

To conclude, the evaluators of a panel (according to topics within a field of knowledge that is established with an eye on a disciplinary field) will have to quantify the amount of experience, feasibility, innovation, interdisciplinarity and relevance that they find in each one proposal, among hundreds or even thousands, from their own perspective, according to their own disciplinary background(s), etc. It is evident that this is no easy task for anyone and it is certainly an enormous responsibility for both sides because there is a lot involved, no matter you look backwards, to your recent professional past, or forward, to your promising short-term future. I shouldn't perhaps say this, to my own interest ... but there is something fictional, let's call it creativity, in the planning of future research. Therefore, the success of your research plan depends a lot in your ability to deal with the unexpected without losing track of the path you are walking, that is, you do not know exactly where you are arriving but, for sure, you cannot allow yourself to get lost on the way, at any point.

With the ALiB and LEFHBio colleagues, I could only be surprised to find out that we were fighting the same struggle through different paths – plurilingualism and intercultural epistemologies – and that I had found what I was hoping for. They were two different "tribes" – to use Latour and Woolgar's (1986) words – two research groups, who welcomed my project with scientific curiosity, with so much knowledge and wisdom in their own fields that they naturally gave me the space for critical interrogation and ourselves for mutual learning, something that might have easily sounded uncomfortable, even impertinent, in other circumstances, with other research groups.

Concerning their responses toward linguistic and cultural hegemony in scientific production, both research groups are so much involved with diversity in Brazil, both linguistic and biological, that the predominance of linguistic and cultural hegemony of English in scientific production, in the name of globalization, sounds too anachronistic and pointless to them. It should not be forgotten that, due to the dimension of this country, the Brazilian academy can feed itself. I am not saying that this is desirable, I am saying that Brazilian scholars can easily survive without English. From my point of view, my project was offered, by these two research groups as well as by the other research groups from the other two universities, a very solid decolonial South-North perspective, which can be perceived and apprehended in the previous chapters of this book. They have their scientific studies well settled in the Brazilian context, with so much diversity at hand, that they are more concerned about responding vigorously to their mission. Hence, their work in the Brazilian context was also very inspiring for the concept of Intercultural Responsibility, which I have been developing, because of their attention to linguistic and cultural diversity within the context they are studying and their sense of civic responsibility in preserving and valuing such linguistic, cultural and biological diversity and, finally, in bringing it to the scientific fore in order to claim for its academic, social, and political validity.

Finally, Manuela, Can You Tell Us a Little about the Interdisciplinary Event that You Decided to Organize during the Bahia Stage of the Project's Development?

This interdisciplinary event was not in my project plan and I organized it only at the *Universidade Federal da Bahia* because the project had been so successful there, both in its first phase, with the language teachers and, in the second phase, with the two research groups mentioned above. First, it was physically feasible because both departments were located in the same campus and very close to each other, although their members had never met with each other yet, at least officially. If we stopped once and reflected about how the daily routines of our disciplinary and disciplined departments develop mechanically within conventional borders, we would ourselves consider unbelievable that we spend our lives, side by side, studying the same realities from different but complementary perspectives without even noticing each other or even knowing about each other, not even greeting each other when we rush inward and outward the buildings. Second, and foremost, I felt that I was dealing with adventurers capable of crossing the invisible lines, with inspiring and inspirable colleagues. And, in fact, they did, punctually, bravely, eagerly, open their office doors, down the stairs, walked across the lawn and up the stairs, and sat down around the only one largest table we could find to accommodate more than twenty people facing each other. Finally, I had been so touched by their generosity and inspired by the exciting experience I was having in crossing the disciplinary lines, that I thought that the only way I could possibly compensate them was to provide them with a similar opportunity to experience, with the same learning intensity, what I had experienced. The participants in this interdisciplinary event were all there in their position as "glocademics", that is, inter- and intra- national researchers, either in Applied Linguistics (Portuguese, English, and Spanish), in a greater number, or in Biology, Philosophy of Science, and Science Education.

This 3-hour long enlarged and stimulating conversation contributed to conceptual discussion and practice sharing, from an interdisciplinary perspective, of the following terms and expressions which were the focus of the Glocademics project: linguistic variation, language power and empowerment, glocal languages, comparative analysis, intercultural epistemologies, intercultural translation, *interculturalidade*, global and local knowledge, universality and locality, inter- and trans-disciplinarity, academic centers and peripheries, scientific writing dominant patterns, the intercultural, epistemological, social and political role of scientific research, and intercultural responsibility in scientific research. The discussion was very much convergent as far as the conceptualization of the abovementioned terminology was concerned, the participants were acquainted with most of the terms and were open to discuss the new terminology introduced by the Glocademics project conceptual matrix, namely glocademics, glocal languages, and intercultural responsibility. I believe that I can say that most participants displayed a South-North scientific perspective, that is from local to global, and they were familiar with the theories of the Epistemologies of the South. However, they could recognize, in general, that not only their theoretical inputs were still predominantly northern based, mainly European and North-American, but they also tended to select conferences in Europe to attend.

This interdisciplinary event celebrated an intensive cooperation of two years that also included my collaboration in the post-graduate program by invitation of its coordinators, Américo, yourself and Denise, who were also participating in this enriching conversation. All these activities, which lasted for 2 years, although shared with the other three universities, created a professional and personal bond that justified this get-together where all participants were engaged in an organized brain-storming session. Three of the five research groups I worked with, were settled in Bahia, the third one at the UFSB, the *Universidade Federal do Sul da Bahia*, as you well know one of the newest universities founded when Lula da Silva was President and set up by a previous Chancellor of UFBA, Naomar de Almeida Filho. Its excellent design closely follows the main Brazilian thinkers of education, Anísio Teixeira, Paulo Freire, Milton Santos, and others such as Portuguese scholar

Boaventura de Sousa Santos and Tunisian-French philosopher, Pierre Levy. The UFSB mainly serves a deprived population in the South of Bahia, primarily indigenous communities who have survived the arrival of Portuguese colonizers and the neglect of Brazilian governments as well as some communities of *quilombolas*, impoverished communities resulting from the congregation of enslaved African who fled from their masters and formed their own communities or joined existing indigenous ones (see chapter 8).

Unfortunately, the best undertakings of education are now moving way backward in Brazil, more particularly with the Bolsonaro government, having already worsened since Dilma Rousseff's second election, in 2014, precisely when I arrived in S. Paulo to start the Glocademics project, and her subsequent impeachment. The research group I worked with at UFSB was attempting to recover, together with the community researchers and teachers, the original *pataxó* language of the tribes in the region, their art and culture. The coordinator of this group is a highly prepared scholar and both this research group and the university were vibrantly starting during my first visits and shockingly fading away when I left in August 2016. Very sad indeed. I would prefer to finish this interview with a positive note, by only reminding all the intellectual wealth that I found in the universities I visited and in the work being carried out by the enthusiastic colleagues with whom I worked, but unfortunately I cannot but regret the dismantling of education and research initiatives that the country education and science institutions have been suffering since I have left and even while I was there working on this project while Dilma Rousseff was struggling with a ghastly impeachment process. The Glocademics project proposal was designed with the inspiration of the thrilling vision for education and science that had been set into place during the previous years and left me with an academic network that I am looking forward to have the opportunity to develop and expand. It is also a good time now to demonstrate our solidarity and, above all, our recognition of value to the scientific achievements of our Brazilian colleagues which have been contributing to the advancement of world knowledge. For, limiting the scope of knowledge validity does not reduce the breadth of our ignorance.

References

Cardoso, S. A. M. et al. Atlas Linguístico do Brasil. Londrina: EDUEL

Guilherme, M. (2018) 'Glocal languages': The 'globalness' and the 'localness' of world languages. In S. Coffey and U. Wingate (eds.) *New Directions for Research in Foreign Language Education* (pp. 79–96). Abingdon: Routledge

Guilherme, M. (2019) Glocal languages beyond postcolonialism: The metaphorical North and the South in the geographical north and south. In M. Guilherme & L. M. T. M. Souza (eds.) *Glocal Languages and Critical Intercultural Awareness: The South Answers Back* (pp. 42–64). London and New York: Routledge

Guilherme, M. (2020a) Intercultural Responsibility: Transnational research and glocal critical citizenship. In J. Jackson (ed.) *The Routledge Handbook of Language and Intercultural Communication* (2nd ed., ch. 21). Abigdon, UK: Routledge

Guilherme, M. (2020b) Intercultural Responsibility: Critical inter-epistemic dialogue and equity for sustainable development. In Leal Filho W., Azul, A.M., Brandli, L., Lange Salvia, A., Wall, T. (eds.) *Partnership for the Goals: Encyclopedia of the UN Sustainable Development Goals*, vol. 17. Springer Nature: Cham, Switzerland

Latour, B. and Woolgar, S. (1986) *Laboratory Life: The Construction of Scientific Facts*. New Jersey: Princeton University Press (2nd Edition)

Machado Filho, A. V. L. and Oliveira, I. P. S. (2017) O português como língua *glocal*: Aspectos sócio-históricos e linguísticos de sua conformação. *Filologia e Linguística Portuguesa*, 19: 2, 257–270

8 Glocademia: Intercultural responsibility across North/South epistemologies

Manuela Guilherme

"Glocademia" is the collective word, hereby introduced (please see Figure 0.1 in the Introduction), to refer to the *ensemble* of "glocademics", academics who develop their activities both globally and locally. This terminology was previously introduced through a project entitled "Glocal Languages and Intercultural Responsibility in a Postcolonial Global Academic World (GLOCADEMICS): Power Relations between Languages/Cultures within and between Research Groups" (http://www.ces.uc.pt/projectos/glocademics). The project was supported via an individual EC-MSC (European Commission - Marie Sklodowska-Curie Actions) three-year grant (2014–2017) whose empirical study was implemented in Brazil for two years and consisted of a "research on research" (Enserink 2018) study with five research groups from three Brazilian federal universities from different (inter)-disciplinary fields both in the social sciences and life sciences. It took its perspective from the humanities regarding language, culture, and epistemology while focusing on their research activities for the production of specific knowledge.

In using the term "glocademics", I am referring to the academic community that responds trans- and intra-nationally to the glocal, simultaneously global and local, to demands of multi-level citizenship engagement while embracing their communities of practice as peer researchers and committing themselves to the recognition of a wealth of knowledge ecologies. Being a transnational researcher – a glocademic – does not mean being nationally, institutionally, or culturally rootless, just as being an expatriate, immigrant, or cosmopolitan citizen does not erase one's multiple ethnic, professional, communitarian, or affective bonds. A transnational researcher does not, one would hope, cease to be a responsible glocal citizen – simultaneously global and local – or an emotional human being. This new professional identity of a transnational researcher – a glocademic – deserves scientific attention beyond that of labor contracts or psychological implications, although these should not be disregarded, by also focusing on reciprocal epistemological rearrangements and mutual learning.

DOI: 10.4324/9781003225812-9

By "glocademia", I refer to the transnational community of academics, both educators and researchers, that today – albeit on occasion for brief periods – move epistemologically across geographies and cultures. They most often do so in order to integrate different teams of researchers and educators and to carry out collaborative educational or research projects as well as develop different professional networks which they later keep through virtual contacts. Such experiences give rise to manifold arrangements of cooperation in which academics are most often not representing a particular nation or home institution but, nevertheless, do not lose their individual ethnic identities, native languages, personal and cultural bonds, particular intercultural strategies, etc. Instead, they keep these sufficiently elastic to carry on with their work keeping in mind the global and transnational context within which academia and citizenry today mingle, while, at the same time and integratively, responding to a local context, which may not be native to them, or even to various localities, when undertaking comparative studies or collaborative and international projects. Both on a professional and personal level, members of academia move in-between spaces, not only by creating "third-spaces", in Homi Bhabha's formulation (1994) but also by articulating multiple spaces and times. It should not be forgotten that these spaces are composed by a multiplicity of languages and cultures and through different ways of being intercultural. Above all, these are inter-epistemic spaces as well.

Science in Society

By the end of the last century, an international group of scholars, identified a "new mode of production of knowledge" that should replace the traditional modernist Mode 1, which they simply entitled as Mode 2, that mainly (1) produces knowledge in a context of application and in the process of application; (2) is transdisciplinary since "the shape of the final solution will normally be beyond that of any single contributing discipline"; (3) is heterogeneous and carried out amidst organizational diversity, for example, "members may then reassemble in different groups involving different people, often in different loci, around different people"; and finally (4) is more concerned about being "socially accountable and reflexive", in that "sensitivity to the impact of the research is built in from the start" (Gibbons et al. 1994). Shortly afterwards, some of these authors expanded their theory by highlighting the dynamics between knowledge production and society and related Mode 2 production of knowledge with Mode 2 society in the "contextualization of knowledge in a new public space, called the agora" and calling for a "socially robust knowledge" (Nowotny et al. 2001: vii). Once Mode 2 society displays more complexity and uncertainty and knowledge production is reshaped by internationalization and globalization, "the continuous construction of the 'local' becomes more important, because it provides the main reference framework in which a sense of stability and orientation can be constructed" (p. 44). Therefore, the authors above acknowledge the need for more reflexivity, creativity and criticality in order to intensify the interaction between science and society where "integration" is replacing "segregation" models and, consequently, altering the researchers' belief systems (Scott 2006, Klein 2010, Barry and Born 2013). Simultaneously, Latour also pointed out "the transition from the culture of 'science' to the culture of 'research'" and he portrays research as "warm, involving and risky" (1998: 208). Without putting scientific rigor at risk, research may add some elements which does not prevent it from also being controversial, ideological, passionate and emotional. Latour also admitted that "there is a philosophy of science, but unfortunately there is no philosophy of research" (ibid.).

Responsible Research and Innovation

The European Commission (EC) politicians and policy makers called for scholars' theoretical support to design a "new" approach of science to societal needs to be implemented under the Horizon 2020 Framework program which came to be entitled "Responsible Research and Innovation". In 2007, in the Lisbon meeting, the Report of the "Science in Society" session (European Commission 2008) points to a "Public Engagement in Science" which proposes to broaden the narrow conceptions of innovation by rooting them in societal needs. In the following year, the EC publishes a report by an expert group (MASIS) entitled "Challenging futures of Science in Society: Emerging trends and cutting-edge issues" (European Commission 2009) where responsible development is geared towards innovation, research governance and technoscience, regarding the nanotechnologies more specifically, and with particular reference to the United States of America research scene. With the European Research Area (ERA), established in 2000, as background, this document also discusses a "European Model of SIS" (Science in Society) where it is claimed that "Europe can be a model, not by arguing for one or more specific approaches, but by **experimenting and sharing** not only the results of experiments but also the processes underlying them" (p. 64, my emphasis). And again the only reference point for comparison/contrast is the US: "Their specificity is often demonstrated by contrasting them [the components of a European model] with the situation and approach in the US" (ibid.). Science production, not to mention knowledge in general, in the "rest" of the world is ignored. Europe is meant to rule in science, experiment, and share, while keeping an eye on the USA scientific state of affairs.

It was in 2013 that the EC issued a report on the "Options for Strengthening Responsible Research and Innovation" (RRI) which attempts to define its ruling criteria, considered as the "processes for [its] successful application" as well as the "instruments to encourage RRI" (p. 4). On the whole, the document examines how the concept of RRI can fit the normative framework of the European research and innovation programs having in mind the backcloth of EU 2020 strategy and the Horizon 2020 program. In the following year, the Rome Declaration on Responsible Research and Innovation was released (Presidency of the Council of the European Union 2014) where RRI is described as "the on-going process of aligning research and innovation to the values, needs and expectations of society". First, that the ways in which technology may be accepted, and perhaps endorsed, by society is one main concern, second, gender inclusion and finally the stakeholders' engagement with regard to science in society, all of which highlight the elements considered fundamental for research and innovation excellence, namely "openness, responsibility and the co-production of knowledge".

From then onward, a corpus of knowledge on RRI has been developed. This theoretical background has moved between two conceptual fields – RRI (Responsible Research and Innovation) and RI (Responsible Innovation) – which are used interchangeably but which I consider far from perfect synonyms as the notion of research places the emphasis on the process whereas its removal in the phrase (RI) puts the product in the center of the innovation process, which is what Innovation is eventually all about. It is therefore meaningful, I believe, that one of the three pillars proposed by the upcoming EC Horizon Europa framework (2021–2027) is precisely "Pillar 3 – Innovative Europe" which entails the creation of the European Innovation Council, European Innovation Ecosystems and the European Institute of Innovation and Technology (https://ec.europa.eu/info/horizon-europe-next-research-and-innovation-framework-programme_en). Von Schomberg, while acknowledging that he takes "a largely European policy perspective" about RRI, "provides a definition of the concept and proposes a broad framework for its implementation under Research and Innovation schemes around the world" (2013: 52). The author gives special attention to the notion of "responsibility", particularly as it implies "collective responsibility" with regard to "the right impacts and outcomes of research", by keeping an eye on technology, not disregarding the stress on and mechanisms for the reassurance of research ethical principles and impact assessment to which the EC had already given particular prominence, mainly since 2002.

According to Stilgoe, Owen and Macnaghten (2013), Von Schomberg's perspective on RRI is "anchored to European policy processes and values" while their view of RI "emerges from a different context" which prompts a "broader definition based on the prospective notion of responsibility described above: Responsible innovation means taking care of the future through collective stewardship of science and innovation in the present" (p. 1570). In addition, they single out four "dimensions" in their RI framework, namely: (a) "anticipation", which implies that researchers are ready to anticipate opportunities for innovation as well as to evaluate the risks involved; (b) "reflexivity", both from actors and institutions, which supports "openness" and prompts "leadership", equally relevant for innovation; (c) "inclusion", pointing to "multi-stakeholders partnerships"; and, last but not the least, "responsiveness", which "involves responding to new knowledge ... [and to] societal challenges", amongst others. This article makes reference to a previous publication by Pellizzoni (2004) who also selects "responsiveness" as one of four dimensions of responsibility, and the one he considers the most neglected feature of all, together with "care, liability, accountability". Nevertheless, Pellizzoni considers responsiveness as the anticipatory capacity of responsibility which requires a readiness to be open to listen to all stakeholders and "to rethink our own problem definition, goals, strategies, and identity" (p. 549–557).

On the whole, I perceive that to be responsible certainly entails being responsive, however, not necessarily the other way round, that is, to be responsive is not all the way synonymous to being responsible. Therefore, I conclude that it is pressing to describe the many possible implications of the RRI approach, that is, responsibility in the context of research and innovation, which I believe needs to be intercultural in itself. Grunwald (2014) examines "techno-futures" within the framework of a "hermeneutic mode", which leads us to focus on the "hermeneutic constellation and to look for an adequate methodology" (p. 282) since, he clarifies, "the objects of RRI debates are not purely scientific-technological developments. These fields are rather being loaded with meanings concerning the future of humans and society" (p. 274). This is also what has led me to introduce and develop the concept of "intercultural responsibility" in the field, as we shall see further below.

Responsible Research and Innovation in Latin America

The idea of Responsible Research and Innovation (RRI) and of Responsible Innovation (RI), perceived as synonyms, is – again – being exported and adapted outside Europe as a new substance in an empty vessel which leaves aside not only the opportunity for critical and creative discussion but also for the possibility for expanding and enrichening this idea. During the last decade, several EC (European Commission) funded FP7 (Framework Program) projects on RRI were developed and year 2019 brought to light relevant publications which can give us support for the work ahead on RRI. For example, Gianni, Person and Reber's edited book (2019) on RRI concepts and practices, offers a comprehensive cartography of the idea of responsibility within the RRI and RI context. The book authors agree on that RRI, and RI, is not a concept, nor should it be, but "an emerging field and social movement", "a policy and political artefact", and that they "are discourses in the making and are interpretively flexible" (Richard Owen's Foreword, p. x). The book authors undertake, across the various chapters, an exegesis of the concept of responsibility and the ways it relates to RRI and RI, while each view stresses its normative, moral, ethical, existential, or pragmatic implications, both for its theory and practice. Armin Grunwald (chapter 2) puts forward "the EEE concept of responsibility" which designates its three dimensions, namely "the empirical, the ethical and the epistemic one" (p. 36). Although "considering all the three dimensions of responsibility together", Grunwald clarifies the relevance of the epistemic dimension in that "it is essential that the status of the available knowledge about the futures to be accounted for is determined and is critically reflected from an epistemological point of view" (p. 38). Bernard Reber (chapter 3) describes RRI as a policy perspective and expands on how it fits the norms and demands which the EC has already set in place. The author considers responsibility to respond to the present time and "intrinsic to action and to the fact that human beings are tied into networks of relationships" (p. 65), which makes him value accountability, responsiveness, and participation.

Innovation governance is also timely discussed by Ludwig and Macnaghten (2020) with regard to Traditional Ecological Knowledge (TEK), who appropriately question the ways in which European scholarship and policy in the field is (not) meeting the governance procedures for innovation in TEK scholarship communities, "characterized as a source for rethinking human relationships with their environments" (p. 1), more particularly those situated in the Global South, both geographically and epistemologically. This is also one of my goals in this chapter and I dare say that it should be perceived as fundamental action in order to reach the accomplishment of the UN 2030 Sustainable Development Goals. In this regard, implementation of RRI in Brazil has been a particular concern of a group of scholars, particularly at the University of Campinas in the S. Paulo region, in collaboration with European experts in the field. These scholars have raised important issues that apply to the global South in this matter. The very first aspect is the idea of "innovation" perceived as "linear progress", to which they refer as "innovationism" (Reyes-Galindo et al, 2019). A prior workshop on "Responsible Innovation and the Governance of Socially Controversial Technologies", at the same university, with "a group of early researchers and academics from São Paulo state and from the UK" is reported to have generated much discussion on the concept of "Responsible Innovation" (RI) which, for example, "one Brazilian participant perceived as intellectual 'neo-colonialization'- that it could unwittingly reproduce or reinforce relations of dependence that are far from emancipatory for the Global South" (Macnaghten et al, 2014: 193). In fact, the idea of "innovation" can easily fall into a new mantra, although tempered by a sense of responsibility, which masks the modernist notion of linear progress imposing "cognitive imperialism". To use Santos' words:

> To a large extent, this is the world built by modern science and the myth—based on the progress of science, economic science included—that all social and political problems will have technical solutions. The myth is still with us, now exacerbated by the revolution in information and communication technologies. Nonetheless, the myth begins to lose credibility
>
> (Santos 2018, p. 160).

Indeed, evidence of climate change, heavy storms, dramatic droughts, fierce fires, nature, and life conditions collapsing does not allow us to passively indulge the "innovation" lullaby as the only way to secure our survival and progress. Throughout his work, Santos, as well as most decolonial scholars such as Mignolo and Walsh (2018) who remind their readers that "the rhetoric of modernity was based on progress and civilization", have unmasked the inexorability of linear time and progress which has always brought us to a point where the same always arrive first while the others remain at the end of the line. Cusicanqui, a Bolivian scholar, explains that "the indigenous world does not conceive of history as linear; the past-future is contained in the present ... the project of indigenous modernity can emerge from the present in a spiral whose movement is a continuous feedback from the past to the future ..." (2019: 107). How do the different perceptions of the present may impact on the idea of innovation? How do the various perceptions of "innovation" and "creativity" differ worldwide?

Among the initial results found by the Brazilian team in the RRI-Practice project (Responsible Research and Innovation in Practice), Reyes-Galindo, Monteiro and Macnaghten (2019) highlight the importance given by the Brazilian project participants, all involved with research or its management, to "multiculturalism and ethnicity, two issues that are underdeveloped within the European RRI framework" (p. 357). Even gender discrimination, a main issue in the European RRI framework and which was often considered a "non-issue" in the Brazilian study, "in the Brazilian setting is intimately tied to racial and social inequality, elements not directly considered in the RRI keys" (ibid.). Nevertheless, although bearing in mind that this study was only carried out in "a very specific region in the country and not representative of Brazil as a whole" (p. 355), the authors had to admit "the overwhelming appeal to the linear model of innovation, in which funding of basic science is seen as leading to technological development and subsequently to successful diffusion into society" (p. 356). However, neither this article, nor other work on RRI mentioned above, mention "university outreach programmes" in Latin American university systems as a traditional model which can provide inspiration for the development of RRI, an element which I found very inspiring for the conceptual framework of Intercultural Responsibility that I was developing. Since RRI, as a term, has not been used in Latin American research policies, the concept is also assumed to be inexistent (e.g. see bibliography above and a video recorded interview of the NewHorrizon project (https://newhorrizon.eu/) entitled "Aspirations of New Horrizon" where RRI is said to be "a new topic in Latin America" (https://www.youtube.com/watch?time_continue=151&v=YmVrlfeKBhY&feature=emb_logo [at min.1.33), in Colombia more specifically.

From "Science in Society" to "Society in Science"

Latin American university life has been officially based on a tripod framework with Teaching, Research and Outreach (*Extensão*) as three equally important pillars, since the beginning of last century. University outreach program have grown more and more influential in the other two components even though the perception of its role in society changed according to the ruling powers and ideologies of each time, that is, by oscillating between welfare (*assistentialismo*) and empowerment, in dictatorship and democracy respectively. The moral and practical duties of higher education toward society date from its creation and its expansion all over the world, and have been reiterated ever since, for example by UNESCO's World Declaration of Higher Education in the 21st Century (2005). However, in Latin America, the model of university outreach programs, which interests us here, for the purpose of the development of Responsible Research and Innovation and of Intercultural Responsibility in research, aims to reflect the social context and work with the community, while envisaging political impact. This model was originated in the 1918 Cordoba Movement/Reform, in Argentina, by students who claimed against the Eurocentric model of the Latin American university model and required that it set its roots in their social context.

In Brazil, the University Outreach programs were officially recognized in 1931 with the regulation of Brazilian universities (*Estatuto das Universidades Brasileiras*) and its role changed with political regimes. It was in 1987, with the country new re-democratisation, that the National Forum of Pro-Rectors of University Outreach at the Brazilian Public Universities (FORPROEX – *Fórum Nacional de Pró-Reitores de Extensão Universitária das Universidades Públicas Brasileiras*) was established, indeed a fundamental step to ensure its future and footprint (Romão 2018). However, at the same time, Freire, one of the main instigators of outreach activities in rural areas, critiqued the term "*extensão*" which he perceived as an one-way flow, that is, an intent to impose academic knowledge on society, that is, "Science in Society". Therefore, he proposed the word "communication" to replace "extensão" through his book entitled precisely "Extensão ou Communicação?", firstly published in Spanish in 1969, while still in exile in Chile (1983). Freire was an apologist of knowledge creation through dialogue and, in this case, according to his own experience, knowledge should develop from the culture in place in the social micro-context, that is "Society in Science" instead. Other scholars are also considered classical galvanizers of university outreach in Brazil, perceived as closely intertwining with teaching and research, namely, Darcy Ribeiro, Florestan Fernandes and Milton Santos (Antunes, Gadotti and Padilla, 2018). Popular universities and intercultural universities (dedicated to indigenous communities) in Latin America represent the essence of extension/communication activities (Santos, Mafra and Romão 2013, Dietz, 2019).

Emblematic Examples of Practice

My research experience in Latin America, as scientific-co-coordinator of a 3-year European funded project on higher education, RIAIPE3, with 22 teams scattered throughout 14 countries in South and Central America (chapter 2) followed by another 3-year project in Brazil, GLOCADEMICS (http://www.ces.uc.pt/projectos/glocademics), as Principal Investigator, carried out in 4 federal universities, provided me with examples of research projects and activities where outreach, research and teaching were intertwined. As research work flows naturally with outreach into the community and teaching, with students involved, I could observe many examples of an integrated way of doing that joins teachers/researchers, students/researchers, and social stakeholders/researchers. These examples though cannot be taken for granted given the huge dimension and diversity of the target population and the institutions involved. Nevertheless, good practices generally display a focus on plurilingual and intercultural epistemologies and have provided me with material for developing the concept of intercultural responsibility from a perspective of the South, both geographical and metaphorical.

An example of universities in Brazil where community engagement has become more evident and crucial in academic life, particularly in the sense of Paulo Freire's understanding of intercultural communication between the societal resources, interests, and needs and university capacity for responding to them, that is, within the development of a perspective of Society in Science, are the new popular universities created and established during the Lula and Dilma governments in the beginning of this century. Among these, are noteworthy, due to their deep incorporation in the social and cultural context, the Federal University for the Integration of Latin American Integration (UNILA), in Foz de Iguazu, the so-called three frontier region between Brazil, Argentina and Paraguay, and the Federal University of Southern Bahia (UFSB), both established in territories still heavily populated by descendants of native Americans and where poverty has been distressing. The latter was also a site to which the Glocademics project led me, by 2016, to visit and work with a research group coordinated by Professor Rosângela Pereira de Tugny when they were starting a 2-year project entitled "*Maxakali-Pataxó* Art, History and Language: Holistic and Intercultural Public Education in the South of Bahia Region" (*Arte, História e Língua Maxakali-Pataxó: Educação Pública Intercultural e Integral da Região Sul da Bahia*). The Maxakali-Pataxó are two neighboring native-American peoples who have inhabited the states of Minas Gerais and Bahia.

However, to start with, this project needs to be contextualized in its new university, territory and population. The UFSB (Federal University of the South of Bahia – *Universidade Federal do Sul da Bahia*) was set up by the end of Dilma Rousseff's Presidency, in 2013, and when I last visited it, in 2016, an overall crisis was already starting, during the Temer Presidency after Rousseff's impeachment. During Lula da Silva Presidency, Fernando Haddad, his Minister of Education, expanded the university network in Brazil through the creation of fourteen new popular universities serving special intercultural goals in socio-economic deprived and border areas. The UFSB was one of them and was designed by a Commission presided by its *protempore* Chancellor Naomar de Almeida Filho who had been the Chancellor of the Federal University of Bahia (UFBA) for two 4-year mandates. There, he implemented several pivotal reforms aimed at the inclusion and success of students from all walks of like, for example, by giving preference to those coming from public schools, in order to respond to the quota system, both social and racial, just introduced by the government. The UFSB serves socio-economically deprived populations in the Southeast of Bahia, however, of great cultural wealth. Therefore, the original university design was very ambitious, combining institutional and curricular management guidelines proved successful in leading universities in Europe and North America and, simultaneously, practical studies that attempt to respond to the needs of the population in the area.

This territory and its native population were the first to host the Portuguese sailors and, in the present, we can only find some communities of the *Pataxó* people scattered around villages in the territory of *Porto Seguro*, meaning safe harbor, as well as communities of *quilombolas*, descending from the communities of native people who gave shelter to fleeing African slaves. Tourism has strongly developed in this area and the building of resorts has been the last argument for expelling native communities from their lands, after historical expulsions and killings by coffee and, later, by cacao farmers. Since 1500 that these communities have been persecuted and decimated, fleeing from village to village and also fighting against each other. Such circumstances naturally resulted in unstable family and individual lives. However, their unimaginable tenacity, and unbelievably their segregation, allowed their survival and that of some cultural remains which some of their members are currently attempting to recover through interdisciplinary ethnographic work with academic value.

The UFSB Guide Plan explicitly refers to leading Brazilian mentors on education namely Anísio Teixeira, Paulo Freire and Milton Santos, besides the Portuguese Boaventura de Sousa Santos and the French Pierre Lévy. Such a challenging project attracted prominent scholars from other high-ranking universities in Brazil who felt galvanized by the visionary program. Amongst them, Professor de Tugny, a prestigious scholar from the University of Minas Gerais, with an international background in Music and work done on the *Maxakali* art and culture which she is now pursuing in Porto Seguro with the *Pataxó* descendants. This latter project included researchers both from the UFSB and other universities (Rio de Janeiro, Minas Gerais, etc.) as well as from the *Pataxó* community. Besides becoming acquainted with the university campus, colleagues and Professor de Tugny's work (e.g. De Tugny 2011) and project, through her exceptional hospitality, she also provided me with the opportunity to share and record a meeting, around three-hour long, which took place at the Jaqueira Indian reservation. In this meeting, together with the Director of the Indian Museum in Rio de Janeiro, Professor de Tugny and other UFSB professors, I had the privilege to listen to and talk with some representatives of the *Pataxó* community at Jaqueira, namely Sineide, Oiti, Anari, Nayara, Juari, Tauá, Danette, etc., some of whom had carried out research on their people's history, art, and language, for their master's theses.

The researchers from the *Pataxó* community told us about their efforts to retrieve their language and culture in a systematic way. The *Pataxó* communities in the area have been historically unstable, persecuted, killed, and also suffering from fights inside the communities, moving from one village to another, to different tribes and regions, often back and forth. Their main exchanges have been with the *Maxakali* in Minas Gerais, a contiguous state to the south, as their ancestors appear to have been part of a greater confederation in these coastal areas sharing common linguistic and cultural features. The Maxakali have managed to keep their language and culture stronger and, therefore, they have been resources for the Pataxó to try to recover their lost language. The community researchers have taken advantage of these contacts and of the older people, most of whom have moved back and forth in different villages, and therefore, they have been reminding, collecting and clarifying the semantics and phonetics of their language, the *Pataxohã*. According to them, life descriptions, through words, phrases and sentences, have mostly been preserved in songs. In Minas Gerais, Professor de Tugny had also collected this musical heritage from the *Maxakali* and published a series of collections of songs (De Tugny 2013, 2014) and she was trying to replicate a similar ethnographic study with the *Pataxó* communities.

The dedication of the UFSB scholars and community researchers and people to this work, attempting to recover and preserve the native peoples' heritage with so little funding and support is noteworthy. This project, comprehending the UFSB scholars and the native communities, responded to my project in that they cooperate among themselves, as academic and community researchers as well as with researchers from abroad and from other universities in the country, by building plurilingual and intercultural epistemological exchanges and recovery.

The community researchers mentioned the video material they have produced, the books, the challenge to find the right orthography for the sounds and to find out about the ways other communities are putting the sounds into writing. They are already concerned about reaching as much parameter uniformity as possible. They also added that this is a demanding process because, on the one hand, they have to comply both with the scientific validation requirements from the university and the Council of Leaders' (*Conselho de Caciques*) agreement for each study. On the other hand, it is an advantage that some participating researchers are also members of the community because otherwise the elder, whom they consider as "live books", would not have expanded on their descriptions and remembrances. They also consider the indigenous school, where they are also taught in Portuguese about the general curriculum, as the center of community life, for knowledge reconstruction and exchange across the generations. They shared with us that they had changed their attitude towards the stories of the elder members, the surrounding woods, their routines, while attempting to combine academic and traditional knowledge, although with some individuals, on both sides, one unfortunately ends up rejecting the other. But they also feel that they are breaking ground, they are writing about something that no one has studied before, and the "regained word" (*palavra retomada*), as they name it, has acquired a greater significance for the *Pataxó* people.

The chapters above, from 3 to 6, represent the other four research groups which also collaborated with the Glocademics project as focal groups, and are also emblematic of academic community engagement, the peer relation between the university and social stakeholders and the com research, outreach programs and teaching. Chapter 3, for example, is about a comparative study of the democratic development of public governance, between Brazil and the UK, aimed to create a common conceptual framework for a further project, and the intercultural negotiation of concepts was more specifically what had made it more interesting for the Glocademics project. This goal led into lines of research such as the underlying democratic institutional designs, new forms of flexible public governance and the relationship between citizens, politicians, and the institutions. The language used between international partners was English, however, it is noteworthy that, in this regard, the coordinator of the British group mentioned that the conceptual discussion was richer in this context, because of the different mental models not only between English- and Portuguese-speaking but rather among researchers on the whole. Accordingly, the need for intercultural conceptual discussion between two different political and social contexts had been an added-value to the study because it had introduced more variety into the possible definitions. According to him, this is unlikely to happen "when working with academics whose first language is English because the concepts are not exactly specified", since they are taken for granted, I should say.

The relevance of specific debates, such as the different interpretations and consequences of institutional flexibility and incompleteness, for the Glocademics project is evident, even more so as we take into account the distinct historical and cultural developments of Brazilian and British political and social systems. For example, the British coordinator had concluded that "partisan politics was much more significant in all aspects of government in Brazil than in the UK" and this was, for example, one of the reasons why they were using "unconscious mental models in interpreting each other's governmental systems". Besides, according to them, it was enriching to discuss "these background assumptions each side had regarding the operation of political and governmental systems". This is where comparative analysis needs to take different languages, cultures and epistemologies into account and be aware of their impact on it.

Chapter 4 reports a small part of the ALiB research group's work on the national Linguistic Atlas of Brazil (ALIB). Their team's work, both intra- and trans-nationally, offered fruitful fieldwork for the development of Glocademics' two leading conceptual axes – glocal languages and intercultural responsibility. It was already in 1952 that the Brazilian government issued a declaration in favor of the elaboration of a Linguistic Atlas of Brazil and this group has been renovated through several decades to carry out a total of 1,100 interviews to informants who have provided time and information for free, and research travelling inside the 8,515,767 km^2 Brazilian territory. This is only one evidence of the commitment of regional research group members and of the responsiveness of local informants, accounting for intra- and inter-cultural responsibility with regard to the recognition of Portuguese-as-a-glocal-language and in order to make its local dialectical versions academically valuable and legitimate. Their findings are precious resources for science and education - Portuguese language teaching and learning as well as for the other research projects on either the Social Sciences or the Life Sciences in Brazil that have, for sure, to deal with the Portuguese language peculiarities at the diatopical, diastratic and diagenerational, multiple, and overlapping, levels. Their ultimate goal has been the preservation of linguistic variation in Brazilian Portuguese.

Their workplan does not include the other languages still existing in Brazil, which survived the fierce persecutions of the Portuguese-only policies along its history, such as (a) the indigenous languages; (b) the African ones, with some remains in the *quilombos*, the communities formed by fleeing slaves who joined some indigenous communities; (c) some dialects brought from Europe which survived in Brazil, for example, the Pomeranian dialect which has disappeared in Germany; (d) and others, such as sign languages and hybridisms in the frontier zones, for example, at Foz de Iguaçu – the three-frontier zone – in the southwest, where Portuguese, Spanish and indigenous languages cohabit, and the same in the northeast, the state of Amapa, where Portuguese, French and indigenous languages coexist, besides the long Amazonian border, yet less populated. However, their project outcomes have given high prominence to the influence of these and other immigrant languages in Brazilian Portuguese. It is also evident that the ALiB linguistic cartographies display the heritage of a specific colonial matrix where miscegenation surpasses segregation and, hence, where it is noticeable a vast shade of Brazilian Portuguese, which is the result of a rather linguistic attitude, permanent hybridization, and adaptation. Noticeable are also scattered pockets of identifiable languages throughout the territory, none of which could endure absolute impermeability, not only because of the overwhelming dominance of the Portuguese language but also because of Brazilian Portuguese extreme elasticity.

Chapter 5 deals with a challenging task which I would call scientific "intercultural translation" (Santos 2018), aiming at extreme accuracy. These natural science experts were challenging the frontiers of knowledge in order to respond to the extreme variety of Brazilian foods and recipes into which they were translating a software originally created in English for European types of gastronomy. The researchers acknowledged that they often ended up solving intuitively the issues raised, but according to their descriptions of the process and the examples provided, I concluded that they had reached solutions after long-lasting and empirical searching, questioning, and reflecting. They clarified that, despite the linguistic and cultural adaptation of the software, its format and the methodology still enabled that comparability and final recommendations could be made possible, which was the ultimate goal of this project. Nevertheless, one researcher cautiously warned that it was likely that some "dietary intakes" of different countries and regions might not be comparable.

After all, researchers stated that they had never felt that there could be any conceptual misunderstanding because, on the one hand, with regard to difference, they exhaustively discussed each item amongst them and with the colleagues from the coordinating team at the home organization who had already implemented the software in English and, therefore, knew the software well. Besides, with regard to what was common, they used a methodological terminology, e.g. variability, feasibility, usability, validity, etc., which is sufficiently discussed in the literature and, therefore, could control their research process. Not only did they have to deal with different products but also with different names for the same product across regions. Some of these products are hardly recognized in other regions due to the size of geographical areas and the large distance between them, within Brazil, which are only recently covered by flights not yet easily affordable by most of the population. Besides, this research group could not ignore economic inequalities between regions and between social classes in Brazil, resulting in malnutrition, more evident among the more fragile, e.g. children and elderly, nor overlook the current nutritional transition, from traditional to western-type food, that is taking place at different rhythms in different places or their relation to chronic diseases. The researchers were aware that they were trying very hard to minimize perception errors in order to meet methodological standardization and, simultaneously, respect Brazilian rich gastronomy. It sounds paradoxical but it is not conflicting, they said.

Chapter 6 describes one of the transnational projects in which the respective research group is building a knowledge base integrating academic scientific knowledge with traditional knowledge from fishing and indigenous communities in the northeast of Brazil and Namibia, together with practical knowledge from environmental technicians and teachers, for use in conservation and education efforts. During a focus group final discussion, the participants mentioned that science has also been subject to preconceived generalizations and, therefore, the perceptions that each scientific act is the result of local and multicultural activities and that sciences are not monolithic have been overlooked. Such a myth has, in its core, been built by those who see themselves as dominant as well as those who have incorporated the idea that they are dominated, as far as science production is concerned. I understood their statement as a clear authorship claim to "Science". Furthermore, this group has been very committed to build an integrated corpus of knowledge together *with* non-academic project stakeholders, without creating a rupture with science.

One of the group members pointed out how important it is to define "culture", "interculturality" and "science" in order to establish the relationships between them. Another focus-group participant confirmed that, nevertheless, science always has a transnational dimension, that is, it is not national *per se*, although individual sources have national origins, single or multiple. However, they acknowledged some "conceptual imperialism", from other languages and scientific canons, in English mainly, understood as global/universal, sometimes national but also of European origin and academically dominant, that they are forced to adopt in order to make their work intelligible to a wider audience and, therefore, published internationally (or even nationally or regionally).

Within inter-university projects, they have also been committed to develop a theory of "conceptual profiles" geared to science education by creating "epistemological matrices" of some predominant concepts in the field of biology education, for example with regard to the Darwinian concept of adaptation (Sepulveda, Mortimer and El-Hani 2013). Their work on conceptual profiles also results helpful for the development of a theory on intercultural epistemologies, although this is not put forward by their authors (Mortimer, Scott and El-Hani 2012, 2014). Conceptualization is understood as socially situated and embodied, not merely as a cognitive activity, involving "the brain, the body and the environment" (Mortimer, Scott, Amaral and El-Hani 2010). In sum, conceptual polysemy in science is an intercultural resource to be taken into account in transnational research, be it in the natural sciences or in the social sciences.

Intercultural Responsibility

The research groups above carry out research activities that are directly involved with society, languages, cultures, and intercultural epistemologies and promote civic responsibility. They seek plurilingual and intercultural knowledge in society that informs scientific fields such as biology (ecology and nutrition), political science, applied linguistic, and cultural studies. Thereafter, while reorganizing such knowledge scientifically in collaboration with social/community stakeholders, as well as with international/transnational research partners, academic researchers share intercultural responsibility with local stakeholders and research participants. Besides Science in Society, there is Society in Science, according to Paulo Freire, not only outreach activities but also knowledge creation through epistemological communication and community engagement. In Freire's words:

> Even if we agreed – which is not the case - with the "extensive" action of knowledge, through which one subject provides it to another (who then ceases to be subject), it should be necessary not only that the signs had the same meaning, but also, but also that the knowledge contents provided were generated in a terrain which was common to both poles of the relation.
>
> (Ainda quando estivéssemos de acordo – o que não é o caso – com a ação "extensiva" do conhecimento, em que um sujeito o leva a outro (que deixa, por isto mesmo, de ser sujeito), seria necessário não sómente que os signos tivessem o mesmo significado, mas também que o conteúdo do conhecimento estendido se gerasse num terreno comum aos pólos da relação.
>
> (Freire 1983: 49, my translation)

In the case study above *with* the *Pataxó* community, and likewise in the ones described in the previous chapters, as well as in the Glocademics project itself, one can recognize the "give-and-take" of knowledge, the creation of an "ecology" of sustainable knowledge (Santos 1999). The case-studies described above and throughout this book were not carried out merely *about* one targeted population or one topic, they co-created *with*. Not that research cannot be both, but that when it envisages "critical intercultural co-responsibility" the *about* is carried out *with* those to whom the study is directly meant, and the latter is prevalent. This process is described as a "metacognitive complicity that brings together the bearers of scientific knowledge and the bearers of artisanal knowledge" and identified as "postabyssal knowledge [which] is always co-knowledge emerging from processes of knowing-with rather than knowing-about", that is, knowing is learning together from the other side of the abyssal line (Santos 2018, p. 147).

Knowledge creation from the other side of the line is not only inter- and trans-disciplinary, but intercultural as well, where generated knowledge is multilateral, that is, common to both (or multiple) cohabiting and cross-fertilizing linguistic and cultural poles, while considered in epistemological equity standards, at different, perhaps conflicting, but translatable levels. Such knowledge created in partnership is meant to improve life in the research groups' communities, in combined terms, namely academic, experiential, artisanal, and artistic, above all based on intercultural responsibility, which makes the exchange reciprocal and balanced between cooperating partners. Intercultural responsibility, as a term and a conceptual framework, emerged more specifically in praxis, from research and practice focused on intercultural competence (Leonardo da Vinci ICOPROMO, https://www.ces.uc.pt/ces/icopromo/) and on plurilingual and intercultural epistemologies (Marie Sklodowska-Curie GLOCADEMICS, http://www.ces.uc.pt/projectos/glocademics).

Intercultural responsibility (IR) is here understood as "substantive", not "formal", neither legal nor moral, according to Jonas' categories (1984), since it does not obey to pre-established and fixed norms, but whose patterns of behavior do not offend legal and moral principles of associated partners nor of national or international laws. Jonas considers that substantive responsibility has been non-reciprocal and vertical, since it responds to a patriarchal model of care (Jonas 1984: 90–98), which is not shared by my conception of intercultural responsibility. On the contrary, I perceive responsibility, that is intercultural, which can also be interreligious, as being reciprocal and horizontal, presupposing intercultural negotiation within a context of every partner's empowerment, although still asymmetric which gives Jonas an argument for its non-reciprocity (1984: 94). Noteworthy is Jonas' main focus on the future which demands a "new ethics of responsibility" considering that human action has strongly changed and is going to change, even more so due to the technological age, on the one hand, and to the increasing "vulnerability of nature", on the other hand. By this, Jonas means "not the *ex-post facto* account for what has been done, but the forward determination of what is to be done" which implies "feeling responsibility" and, therefore, also incorporates "substantive responsibility". Yet, he didn't go as far as he could with regard to countering the patriarchal pattern, according to Bernstein: - "Why not recognize – contra Jonas – that there is *not* a single paradigm or archetype of responsibility?" (Bernstein 1994: 845).

Jonas was among a group of post-Second World War scholars, of Jewish origin, who were deeply concerned about the notion of "responsibility" which they addressed from different perspectives, yet all of them grounded on cultural diversity, such as Arendt and Levinas, as well as Weil and, later, Derrida, etc. While Levinas (1985, 1991, 1998) highlights individual responsibility for an Other, Jonas and Arendt (1951, 1984, 2003) deal with collective responsibility, although the latter was critical of "collective guilt" (Arendt 2003: 21; Bernstein 2002: 223).

Arendt's *leitmotifs*, in her appreciation of responsibility, can be summed up in her considerations about judgment, action and membership, all of them important elements for questioning cultural pluralism. She examines responsibility in the context of the genocide during the holocaust in Europe, however, her assumptions can be translated to current and past situations of physical, linguistic and cultural slaughter around the world (1951). She evaluated the various kinds of relationships between judgement, thinking and knowledge (1984), between action and power and the possibilities for collective resistance, which can determine individual and community actions (2003). Arendt is also concerned about the stateless people for whom no one claims, deprived of human rights, having no right of opinion or to action, and interrogates herself about peoples' paralysis, when even in collaboration, towards committing evil deeds upon each other when the occasion comes, for one reason or another (1951). Arendt perceives the human being as ontologically plural and, together with Jonas, she feels disquiet about the uncertainty of peoples' responses to unexpected situations to which previous patterns, norms and values do not fit, which can reach extreme generosity as well as go beyond careless irresponsibility to inconceivable evil.

Not only is responsibility to be entrusted to individuals and communities but also to state institutions. The case study with the *Pataxó* community described above, and likewise the ones described in the chapters above, both in the field of education and science, provide examples of intercultural responsibility once individual researchers and research groups are devoting their work to contribute, in partnerships with local stakeholders, to render visible, legitimate and scientifically valuable those languages or dialects, cultures and epistemologies which were previously not considered as such. In addition, they listen to each other about both the requirements of academic work and scientific rigor and the communities' traditional knowledge, needs, and options. Intercultural responsibility viewed from a critical, intercultural and decolonial perspective is supported by a recent but prolific theoretical corpus developed around a Philosophy of Liberation, the Epistemologies of the South and the Decolonial Turn, which are complementary and reinforce each other. Dussel (1998, 2004, 2008) explains that the Philosophy of Liberation deviates from Levinas' Responsibility for the Other because it negates the exteriority of the Other. It departs from the local realities, in Latin America and beyond, without aiming to make their principles abstractly "universal". It nevertheless feels the need to share experiences and exchange knowledge in order to strengthen the links between the Global South, so that it turns possible to initiate the South-North relations in different terms.

Likewise, the Epistemologies of South, a geographical metaphor for situated epistemologies, which are materializing and confronting the Epistemologies of the North, both in the South and in the North, claim for "cognitive justice", for "intercultural translation" and for an "ecology of knowledges" (Santos 2014, 2018). From the Global South, the calls for a Decolonial Turn demand a "thinking otherwise" that generates social, political and epistemological critique and radical systemic change in the geographical South leaderships to be grounded in activism and community engagement (Castro-Gómez 2005, Castro-Gómez and Grosfoguel 2007, Walsh 2007, Mignolo and Walsh 2018).

Responsibility gained a colonial and patriarchal code of care whose knots decolonial theories mentioned above have tended to undo as well as community academic practices described in the previous chapters. Therefore, intercultural responsibility aims to very closely question power relations, social asymmetries and transformative actions confronting linguistic, cultural and epistemological unfairness (Guilherme 2020a, 2020b). Trans- and intra-national, plurilingual, intercultural and inter-epistemological research partnerships can be effective strategies to provide sustainable knowledge that counters fear, suspicion and the scientific "Hubris of the Zero Point" (Castro-Gómez 2005). In the post-COVID 19 pandemic and climate change era, transnational research will need to be more dialogical, rather than aiming to expand only, by disseminating globally what is unilaterally generated in a dominant center and, therefore, not effective in contexts to which it is alien. Intercultural responsibility cannot but be addressed critically from different intercultural stands and be meant for mutual accountability.

References

Antunes, A. B., Gadotti, M. and Padilla, P. R. (2018). Reinventar a universidade a partir da extensão universitária. In M. Carnoy and M. Gadotti (eds.) *Reinventando Freire, reinventando a educação* (pp. 209–225). Sao Paulo: Instituto Paulo Freire, Stanford: University of Stanford/Lemman Center

Arendt, H. (1951). *The Origins of Totalitarianism*. New York: Meridian Books, The World Publishing Company.

Arendt, H. (1984). Thinking and moral considerations. *Social Research*, 5:1/2, 7–37

Arendt, H. (2003). *Responsibility and Judgement*. New York: Schocken Books

Barry, A. and Born, G. (2013). Interdisciplinarity: Reconfigurations of the social and natural sciences. In A. Barry and G. Born (eds.) *Interdisciplinarity: Reconfigurations of the Social and Natural Sciences* (pp. 1–56). Milton Park: Routledge

Bernstein, R. J. (1994). Rethinking responsibility. *Social Research*, 61:4, 833–852

Bernstein, R. J. (2002). *Radical Evil: A Philosophical Interrogation*. Cambridge: Polity

Bhabha, H. (1994). *The Location of Culture*. London and New York: Routledge

Castro-Gómez, S. (eds.) (2005). *La Hybris del Punto Cero: Ciencia, raza e ilustración en la nueva Granada*. Bogotá: Pontificia Universidad Javeriana

Castro-Gómez, S. and Grosfoguel, R. (2007). *El Giro Decolonial: Reflexiones para una diversidade epistémica más allá del capitalismo global*. Bogotá: Siglo del Hombre Editores

Cusicanqui, S. R. (2019). Ch'ixinakax utxiwa: A reflection on the practices and discourses of decolonization. *Language, Culture and Society*, 1:1, 106–119

De Tugny, R. P. (2011). *Escuta e Poder na Estética Tikmũ'ũn Maxakali*. Rio de Janeiro: Museu do Índio, Funai

De Tugny, R. P. (ed.) (2013). *Cantos dos Povos Morcego e Hemex-espíritos: Narradores, escritores e ilustradores tikmũ'ũn da Terra Indígena do Pradinho*. Belo Horizonte: Universidade Federal de Minas Gerais

De Tugny, R. P. (ed.) (2014). *Cantos do Povo Gavião-Espírito: Narradores, escritores e ilustradores Tikmũ'ũn da Terra Indígena de Água Boa*. Rio de Janeiro: Museu do Índio, Funai

Dietz, G. (2019). Multiculturalism, indigenism and new universities in Mexico: A case of intercultural multilateralities. *Journal of Multicultural Discourses*, 14:1-2, 29–45

Dussel, E. (1998). *Ética de la Liberación en la Edad de la Globalización y de la Exclusión*. Madrid: Editorial Trotta

Dussel, E. (2004). Ética de la Liberación: Hacia el 'punto de partida' como ejercicio de la 'razón ética originaria'. In K-O Apel and E. Dussel (eds.) *Ética del Discurso y Ética de la Liberación* (pp. 269–289). Madrid: Editorial Trotta

Dussel, E. (2008). Philosophy of liberation, the postmodern debate, and Latin American Studies. In M. Moraña, E. Dussel and C. A. Jáuregui (eds.) *Coloniality at Large: Latin America and the Postcolonial Debate* (pp. 335–349). Durham: Duke University Press

Enserink, M. (2018). Research on research. *Science*, 361:6408, 1178–1179

European Commission (2008). *Public Engagement in Science*. Brussels: Directorate-general for Research (https://ec.europa.eu/research/swafs/pdf/pub_other/public-engagement-081002_en.pdf)

European Commission (2009). *Challenging Futures of Science in Society: Emerging Trends and Cutting-Edge Issues*. Brussels: Directorate-general for Research (http://www.securepart.eu/download/com-2009_masis_report_expert-group_en150625092421.pdf)

European Commission (2013). *Options for Strengthening Responsible Research and Innovation: Report of the Expert Group on the State of the Art in Europe on Responsible Research and Innovation.* Brussels: Directorate-General for Research and Innovation (https://ec.europa.eu/research/science-society/document_library/pdf_06/options-for-strengthening_en.pdf)

Freire, P. (1983). *Extensão ou Comunicação?* Rio de Janeiro: Paz e Terra (1st edition in Spanish, 1969, Santiago do Chile: Instituto de Capacitación e Investigación en Reforma Agrária)

Gianni, R., Pearson, J. and Reber, B. (eds.) (2019). *Responsible Research and Innovation: From Concepts to Practices.* Abingdon: Routledge

Gibbons, M., Limoges, C., Nowotny, H., Schwartzman, S. Scott, P. and Trow, M. (1994). *The New Production of Knowledge: The Dynamics of Science and Research in Contemporary Societies.* London: SAGE

Grunwald, A. (2014). The hermeneutic side of responsible research and innovation. *Journal of Responsible Innovation*, 1:3, 274–291

Guilherme, M. (2020a). Intercultural responsibility: Critical inter-epistemic dialogue and equity for sustainable development. In Leal Filho W., Azul, A.M., Brandli, L., Lange Salvia, A., Wall, T. (eds.) *Partnership for the Goals: Encyclopedia of the UN Sustainable Development Goals*, vol. 17. Cham, Switzerland: Springer Nature. (https://link.springer.com/referenceworkentry/10.1007/978-3-319-71067-9_75-1)

Guilherme, M. (2020b). Intercultural responsibility: Transnational research and glocal critical citizenship. In J. Jackson (ed.) *The Routledge Handbook of Language and Intercultural Communication* (2nd ed., ch. 21). Abingdon, UK: Routledge

Jonas, H. (1984). *The Imperative of Responsibility: In Search of an Ethics for the Technological Age.* Chicago: Chicago University Press

Klein, J. T. (2010). A taxonomy of interdisciplinarity. In R. Frodeman, J. T. Klein, C. Mitcham & J. B. Holbrook (eds.) *The Oxford Handbook of Interdisciplinarity* (pp. 15–30). Oxford: Oxford University Press

Latour, B. (1998). From the world of science to the world of research? *Science*, 280:5361, 208–209

Levinas, E. (1985). *Ethics and Infinity: Conversations with Philippe Nemo.* Pittsburgh: Duquesne University Press

Levinas, E. (1991). *Otherwise than Being or beyond Essence.* Dordrecht, Netherland: Kluwer Academic Publishers

Levinas, E. (1998). *On Thinking-of-the-Other: Entre Nous.* New York: Columbia University Press

Ludwig, D. and Macnaghten, P. (2020) Traditional ecological knowledge in innovation governance: a framework for responsible and just innovation, *Journal of Responsible Innovation*, 7:1, 26–44

Macnaghten, P., Owen, R., Stilgoe, J., Azevedo, A., De Campos, A., Chilvers, J., Dagnino, R., Di Giulio, G., Frow, E., Garvey, B., Groves, C., Hartley, S., Knobel, M., Kobayashi, Lehtonen, M., Lezaun, J., Mello, L., Monteiro, M., Pamplona da Costa, J., Rigolin, C., Rondani, B., Staykova, M., Taddei, R., Till, C., Tyfield, D., Wilford, S. and Velho, L. (2014). Responsible innovation across borders: Tensions, paradoxes and possibilities. *Journal of Responsible Innovation*, 1:2, 191–199

Mignolo, W. D. and Walsh C. E. (2018). *On Decoloniality: Concepts, Analytics and Praxis*. Durham and London: Duke University Press

Mortimer, E. F., Scott, P., Amaral, E. R. and El-Hani, C. N. (2010). Modeling modes of thinking and speaking with conceptual profiles. In S. D. J. Pena (ed.) *Themes in Transdisciplinary Research* (pp. 105–139). Belo Horizonte, MG: UFMG

Mortimer, E. F., Scott, P. and El-Hani, C. N. (2012). The heterogeneity of discourse in science classrooms: The conceptual profile approach. In B. Fraser, K. Tobin and C. McRobbie (Eds.). *Second International Handbook of Science Education* (vol. 1, pp. 231–246). Dordrecht, Heidelberg: Springer

Mortimer, E. F. and El-Hani, C. (2014). Preface and Introduction. In E. F. Mortimer and C. N. El-Hani (eds.) *Conceptual Profiles: A Theory of Teaching and Learning Scientific Concepts*. Dordrecht, Heidelberg: Springer

Nowotny, H., Scott, P and Gibbons, M. (2001). *Re-thinking Science: Knowledge and the Public in an Age of Certainty*. Cambridge: Polity

Pellizzoni, L. (2004). Responsibility and environment governance, environmental politics. *Environmental Politics*, 13:3, 541–565

Presidency of the Council of the European Union (2014). *Rome Declaration on Responsible Research and Innovation in Europe*. Brussels: Council of the European Union (https://ec.europa.eu/research/swafs/pdf/rome_declaration_RRI_final_21_November.pdf)

Reyes-Galindo, L., Monteiro, M. and Macnaghten, P. (2019). 'Opening up' silence policy: Engaging with RRI in Brazil. *Journal of Responsible Innovation*, 6:3, 353–360

Romão, J. E. (2018). Paulo Freire e a Extensão Universitária. In M. Carnoy and M. Gadotti (eds.) *Reinventando Freire, reinventando a educação* (pp. 189–207). Sao Paulo: Instituto Paulo Freire, Stanford: University of Stanford/Lemman Center

Santos, B. de S. (1999). Towards a multicultural conception of human rights. In S. Lash and M. Featherstone (eds.) *Spaces of Culture: City, Nation, World*. London: SAGE

Santos, B. de S. (2014). *Epistemologies of the South*. Boulder: Paradigm

Santos, B. de S. (2018). *The End of the Cognitive Empire: The Coming of Age of Epistemologies of the South*. Durham: Duke University Press

Santos, E., Mafra, J. F. and Romão, J. E. (eds.) (2013). *Universidade Popular: Teorias, práticas e perspectivas*. Sao Paulo: Liber Livro

Scott, P. (2006). The research revolution and its impact on the European University. In G. Neave, K. Blückert and T. Nyborn (eds.) *The European Research University: An Historical Parenthesis?* (pp. 129–143). New York: Palgrave

Sepulveda, C., Mortimer E. F. and El-Hani C. N. (2013). Construção de um perfil conceitual de adaptação: Implicações metodológicas para o programa de pesquisa sobre perfis conceituais e o ensino de evolução. *Investigações em Ensino de Ciências*, 18:2, 439–479

Stilgoe, J., Owen, R. and Macnaghten, P. (2013). Developing a framework for responsible innovation. *Research Policy*, 42: 1568–1580

Walsh, C. (2007). Shifting the geopolitics of critical knowledge. *Cultural Studies*, 21:2-3, 224–239

Von Schomberg, R. (2013). A vision of responsible innovation. In R. Owen, J. R. Bessant and M. Heintz (eds.) *Responsible Innovation: Managing the Responsible Emergence of Science and Innovation in Society* (pp. 51–74). London: John Wiley

Conclusions: The future of glocal and interculturally responsible academia

The title of the book, *A Framework for Critical Transnational Research: Advancing Plurilingual, Intercultural, and Inter-epistemic Collaboration in the Academy* defines "Research" and "Academy" as the topics in discussion, situating the argument at the confluence of higher education institutions, societal contexts and knowledge construction. Only one verb, "Advancing" indicates the action proposed, meaning looking and moving forward. It aims to anticipate how to proceed, by walking the walk ahead, in the midst of languages, cultures and epistemologies. Accordingly, the adjectives are composite, expressing complex dynamics between a variety of languages, cultures, epistemologies which, to some extent, burst out national borders, as they were formally established and organized inter-nationally, into the trans-national stage. As a consequence, the leading adjective "Critical" cannot but be questioned in face of the whole context and include new world visions, conceptual frameworks and strategic goals. Such a background for research and the academy is not wholly new, the elements have always been there, although they are now much more intense and impossible to ignore. The upcoming challenges and changes are being introduced by the fact that the actors have become critically and interculturally aware of the involving landscape, the potential it offers and the urgency of inter-epistemic transnational collaboration for our survival.

The Glocademia Experience

Having the "Glocademics project" as a backdrop, this book is meant to provide a bottom-up view of heterogeneous globalization, offer a coherent look from the local at the global, display voices from the South (labelled as such) who know about the North as well as voices from the North (labelled as such) who know about the South, both of whom want to learn about the local needs and interests through local-global conversations. This book aims to give the stage to different perspectives taken from various knowledge-making experiences and lay out the pluri-, inter-, and trans- relational types of some of the elements which

DOI: 10.4324/9781003225812-10

have become conspicuously distinct, and simultaneously inter-relational, in local-global dialectics.

The global and the local have up until now been perceived as two separate levels, coexisting while still remaining worlds apart. According to the dominant perspective, both the global and the universal have possessed a natural propensity for invading the local and the individual, that is, the macro flowing into and suffocating the micro. That they might be intertwined, mutually negotiable, reciprocal in influence and valued on equal terms, has not been widely taken into account, despite some voices laying claim to this. For example, Robertson (1995) revived the term "glocalization" whose concept had reportedly been accepted as common sense in Asia, easily understandable given that Asian countries do not have a common history of extensive colonization by European outsiders as happened elsewhere, with the concept of empire being replaced by that of globalization. Urry (2005) explains the existing relationship between the global and the local in Marxist terms, that is, similar to the relationship between the superstructure and the agents, although he finds no linearity between them, instead citing "global complexity". Santos refers to the complex interaction between the global and the local as "the turbulence of scales" which "manifests itself through a chaotic confusion of scale among phenomena" (2014: 82). Canagarajah seeks to clarify this process by stating that "treating the local and global as dichotomies is to misunderstand how the global permeates the local and the local finds expression in the global" (2013: 204). We may conclude that the dynamics existing between the local and the global, in that order, is becoming more and more evident. My position is not about "using the term 'glocal' in order to clarify our understanding of the global dimension, instead, we believe that *the impact of the 'local' on the 'global' is as strong as the reverse and, moreover, they are not in a dichotomous relationship but closely intertwining with each other*" (Guilherme and Menezes de Souza 2019, p. 5).

In fact, the local has been understood, in general terms, as representing those upon whom the global has been imposed, while the global applies to those whose local has been induced globally. Yet, there are increasingly significant signs of the local "talking back" to the global and, therefore, generating a greater balance in this relationship. The World Social Forum, founded and held in Porto Alegre in Brazil, deep within the southern hemisphere, its origins dating back to the beginning of this century, aims to counteract the World Economic Forum, held in Davos, and serves as a good example of the moves toward counter-hegemonic globalization (Santos 2006a and 2006b). The COVID-19 pandemic has forcefully brought into the limelight this reverse perception of glocalization "in which people live far more local lives than in recent decades but with greater global awareness" (Goffman 2020: 48). The same can be said of global warming, a global problem in need of localized solutions and specific multifarious actions. Knowledge-producing workers are particularly well positioned to explore this new epistemological approach for the benefit of the planet and its inhabitants. The COVID-19 pandemic is once more bringing to light the (inter)dependence of the macro- on the micro-context, of the macro-economy with respect to the micro-economy, of the macro-policies in relation to the micro-policies, the prevalence of the state at the expense of the nation. The nation–state seems to be bargaining mainly at trans-national and intra-national levels and the impact on the reorganizing of knowledge/science will prove to be extensive and intensive.

This book, as well as the above mentioned Glocademics project from which it emerges but to which it is not limited, applies a definition of science which goes beyond the natural sciences and which is profoundly touched by its plurilingual, intercultural, and interdisciplinary character, since it extends, in one way or another, into an inter/trans-national dimension. Although the construction of science is nevertheless deeply context-dependent, the impact of its pluri-lingual, inter-cultural, interdisciplinary, and inter-epistemic composition ultimately defines it, which is the argument of this book. However, the definition of scientific methods is rendered independent of the language in which science is produced. Instead, the rationale behind the application of scientific methods relies mainly on academic cultures, e.g. positivism *versus* anti-positivism, which are transversal to geographies, languages and ethnic cultures, that is, scientific parameters of knowledge production possessing the ability to coincide and diverge across the same academic language and culture.

Since the 18th and 19th centuries, with the advent of Illuminism and Scientific Revolution - the epistemological division between the natural sciences and humanities has also been reflected in the physical division within the university *campi*, described in the following terms by C. P. Snow, both scientist and humanist:

> For constantly I felt I was moving among two groups - comparable in intelligence. Identical in race, not grossly different in social origin, earning about the same incomes, who had almost ceased to communicate at all, who in intellectual, moral and psychological climate had so little in common that instead of going from Burlington House or South Kensington to Chelsea, one might have crossed an ocean.
> (1998, p. 2, 1st edition 1959)

Such a physical separation between the Natural Sciences, on one side, and the Social Sciences and Humanities, on another side, is a heritage still found in the most traditional universities, for example, also at the University of São Paulo and at the Federal University of Bahia. At the latter, the Glocademics project participants had never met, although the buildings were very close to each other. It is a privilege to be able to include the views and expertise of these researchers, Glocademics, in applied linguistics, biology and ecology, political science, nutrition, art and ethnographic studies in the same book, targeting the same focus which is after all common to their research routines. Through his famous speech at the University of Cambridge Senate House in 1959, entitled *The Two Cultures*, C. P. Snow reignited a controversy which has recurrently roused many other scholars in different languages, including, amongst others, Santos (1988) and Latour (2013). Despite all the criticism, Snow nevertheless claimed that "with good fortune, however, we can educate a large proportion of our better minds so that they are not ignorant of imaginative experience, both in the arts and in science, ..." (ibid, p. 100). It is sad to have to admit that four decades later, and with the dawn of another century, so little has changed besides rhetoric. The same can be said of civil rights and anti-colonial movements, operating along a similar line and whose struggles remain unresolved throughout the world.

This leads us to considering an "interdisciplinary" approach to research which means that not only do the participating research groups carry out their research activities in different scientific fields, which are to a greater or lesser extent also interdisciplinary, either in the social sciences, the humanities or in the life sciences, but also that our meta-reflexive experiment focusing upon their plurilingual, intercultural and inter-epistemic strategies is itself interdisciplinary (see the Glocademia matrix, Figure 0.1 in Introduction). This undertaking is not simply "multidisciplinary" as it does not deal with disciplinary knowledge separately, nor is it "transdisciplinary" as it does not aim to extend to a field of knowledge that remains above disciplines.

Modern knowledge of European origin was organized, for purposes of specialization, in almost hermetic disciplinary fields and it was so exported to the colonies and endorsed by the elites. However, an "ecology of knowledges" requires that other designs for knowledge organization and management are considered, albeit to a greater or lesser extent. Throughout the past decade, literature on inter- and trans-disciplinarity has bloomed, displaying taxonomies of various types, modes and models. Despite the tensions between the notion of taxonomy and the dynamics of research and knowledge production, which however cannot dispense coherence (Rafols and Meyer, 2010), several definitions of inter- and trans-disciplinarity are being provided that rely on the levels and intensity of disciplinary "integration" as opposed to previous disciplinary "segregation", distinguishing between one and the other (Klein 2010, Angus 2011, Barry and Born 2013). Philosophically speaking, this scope for integration points to the "fusion of horizons" theorized by Gadamer for whom "understanding, rather, is always the fusion of these horizons which we imagine to exist by themselves" (1975: 273). Latour (2013), in a satirical description of an anthropologist's adventure in "reconstituting the value system of 'Western societies'", explains how complicated it becomes "because the Moderns present themselves to her in the form of DOMAINS ...". Furthermore, "she is strongly advised, moreover, to restrict herself to a single domain" (pp. 28–29). Finally, we are invited to imagine that, by chance, she "comes across the notion of NETWORK" and concludes that "there are indeed networks that associate ... elements of practice that are borrowed from all the old domains and redistributed in a different way each time" (pp. 30–31).

With respect to implementation, studies have increasingly focused their attention on the operationalization of inter- and trans-disciplinarity in research projects and developed concepts such as "cognitive convergence/divergence", "epistemic dependence/independence" and the "interlocking expertise" in knowledge integration (Andersen and Wagenknecht, 2013; Wagenknecht, 2015; Andersen, 2016). Likewise, Brydon argues that "interdisciplinarity, internationalization, globalization, and cognitive justice need to be thought of – and addressed – together" (2011: 98). In addition, Burawoy stated: "We need to rethink the social sciences, not from the top down but from the ground up, rooting them in the multiple contexts of their production" (2007: 138). All in all, a response to this claim can be found in Santos' work on the "Epistemologies of the South" and, more particularly, in his proposal for an "ecology of knowledges", of a university that looks more like a "pluriversity", claiming for "cognitive justice" (2014, 2018). The above views on interdisciplinarity are extremely helpful for an understanding and attainment of a critical and decolonial glocademia for which the disciplinary structure of knowledge, although an important resource which cannot be dismissed, is not the be-all end-all. This is also a discussion which is both implicit and explicit in all the chapters above and one that permeates epistemological debates about the Epistemologies of the South as well as about the Decolonial Turn in knowledge construction *with* the Global South.

This book aims to provide different accounts of research projects which provide rich groundwork for critical discussion about the above mentioned theoretical developments and recommended policies, from page 1, namely the trans-nationalization of research, the shift from science to research, the inter- and trans-disciplinary approaches to research, the opening up of science to society, in the name of Science for Society as well as Society in Science, and the commitment to responsible science and innovation within the scope of community engagement. All the chapters above are proof of the potential for a newly expanded field of knowledge production which has deserved deeper critical and metareflexive "research on research" on interdisciplinary, plurilingual, intercultural, and inter-epistemic exchanges. However, despite the immense amount of funding awarded to research activities which are, in fact, transnational, interdisciplinary, plurilingual, intercultural, and longing to be more inter-epistemic in nature, it is evident the lack of awareness from (trans)national funding agencies in this direction.

References

Andersen, H. (2016) *Collaboration, Interdisciplinarity, and the Epistemology of Contemporary Science: Studies in History and Philosophy of Science Part A*, 56, 1–10

Andersen, H. & Wagenknecht, S. (2013) Epistemic dependence in interdisciplinary groups. *Synthese*, 190, 1881–1898

Angus, I. (2011) The telos of the good life: Reflections on interdisciplinarity and models of knowledge. In R. Foshay (ed.) *Valences of Interdisciplinarity: Theory, Practice, Pedagogy* (pp. 47–71). Edmonton: AU Press, Athabasca University

Barry, A. & Born, G. (2013). Interdisciplinarity: Reconfigurations of the social and natural sciences. In A. Barry and G. Born (eds.) *Interdisciplinarity: Reconfigurations of the Social and Natural Sciences* (pp. 1–56). Milton Park: Routledge

Brydon, D. (2011) Globalization and higher education: Working toward cognitive justice. In R. Foshay (ed.) *Valences of Interdisciplinarity: Theory, Practice, Pedagogy* (pp. 97–119). Edmonton: AU Press, Athabasca University

Burawoy, M. (2007) Open the social sciences: To whom and for what? *Portuguese Journal of Social Sciences*, 6: 3, 137–146

Canagarajah, S. (2013) Agency and power in intercultural communication: Negotiating English in translocal spaces. *Language and Intercultural Communication*, 13: 2, 202–204

Gadamer, H-G (1975) *Truth and Method*. London: Sheed & Ward

Goffman, E. (2020) In the wake of COVID-19, is glocalization our sustainability future? *Sustainability: Science, Practice and Policy*, 16: 1, 48–52

Guilherme, M. & Menezes de Souza, L. M. T. (2019) Introduction: Glocal languages, the South answering back. In M. Guilherme & L. M. T. M. Souza (eds.) *Glocal Languages and Critical Intercultural Awareness: The South Answers Back* (pp. 1–13). London and New York: Routledge

Klein, J. T. (2010) A taxonomy of interdisciplinarity. In R. Frodeman, J. T. Klein, C. Mitcham & J. B. Holbrook (eds.) *The Oxford Handbook of Interdisciplinarity* (pp. 15–30). Oxford: Oxford University Press

Latour, B. (2013) *An Inquiry into Modes of Existence: An Anthropology of the Moderns*. Cambridge, MA: Harvard University Press

Rafols, I. & Meyer, M. (2010) Diversity and network coherence as indicators of interdisciplinarity: Case studies in bionanoscience. *Scientometrics*, 82: 263–287

Robertson, R. (1995) Glocalization: Time-space and homogeneity-heterogeneity. In M. Featherstone, S. Lash and R. Robertson (eds.) *Global Modernities* (pp. 25–44). London: SAGE

Santos, B. de Sousa (1988). *Um Discurso sobre as Ciências*. Porto: Afrontamento

Santos, B. de Sousa (2006a) Globalizations. *Theory, Culture & Society*, 23: 2–3, 393–399

Santos, B. de Sousa (2006b) *The Rise of the Global Left. The World Social Forum and Beyond*. Londres: Zed Books

Santos, B. S. (2014) *Epistemologies of the South*. Boulder: Paradigm

Snow, C. P. (1998) *The Two Cultures*. Cambridge: Cambridge University Press (1st ed. 1959)

Urry, J. (2005) The complexities of the global. *Theory, Culture & Society*, 22: 5, 235–254

Wagenknecht, S. (2015) Facing the incompleteness of epistemic trust: Managing dependence in scientific practice. *Social Epistemology*, 29: 2, 160–184

Index

Note: Page references in *italics* denotes figures and with "n" denotes endnotes.

For Product Safety Concerns and Information please contact our EU
representative GPSR@taylorandfrancis.com
Taylor & Francis Verlag GmbH, Kaufingerstraße 24, 80331 München, Germany

www.ingramcontent.com/pod-product-compliance
Lightning Source LLC
Chambersburg PA
CBHW060307220326
41598CB00027B/4258
9781032127064